地方应用型本科教学内涵建设成果系列丛书

生物化学简明双语教程

CONCISE BILINGUAL COURSE OF BIOCHEMISTRY

编 者 陈梦玲 崔竹梅 翟 春

南京大学出版社

图书在版编目(CIP)数据

生物化学简明双语教程：汉、英 / 陈梦玲，崔竹梅，
翟春编. — 南京：南京大学出版社，2016.12(2021.10重印)
(地方应用型本科教学内涵建设成果系列丛书)
ISBN 978-7-305-17956-3

Ⅰ.①生… Ⅱ.①陈… ②崔… ③翟… Ⅲ.①生物化学－双语教学－高等学校－教材－汉、英 Ⅳ.①Q5

中国版本图书馆CIP数据核字(2016)第298523号

出版发行	南京大学出版社
社　　址	南京市汉口路22号　　邮编　210093
出 版 人	金鑫荣

丛 书 名	地方应用型本科教学内涵建设成果系列丛书
书　　名	生物化学简明双语教程
编　者	陈梦玲　崔竹梅　翟　春
责任编辑	刘　飞　蔡文彬　　编辑热线　025-83592146
照　　排	南京开卷文化传媒有限公司
印　　刷	广东虎彩云印刷有限公司
开　　本	787×1092　1/16　印张 18.75　字数 433 千
版　　次	2016年12月第1版　2021年10月第2次印刷
ISBN	978-7-305-17956-3
定　　价	59.00元

网　　址：http://www.njupco.com
官方微博：http://weibo.com/njupco
微信服务号：njuyuexue
销售咨询热线：(025)83594756

PPT：生物化学
简明双语教程

＊版权所有，侵权必究
＊凡购买南大版图书，如有印装质量问题，请与所购
　图书销售部门联系调换

前　言

　　全球经济一体化、国际化的快速推进以及我国复合型人才培养的要求对高等教育产生了巨大影响。教育部2001年发布《关于加强高等学校本科教学工作提高教学质量的若干意见》(4号文件)指出：逐步推广和实行本科生教育中公共基础课和专业必修课的双语教学，是培养新时期我国复合型创新人才的重要基础。2007年教育部启动的"高等学校教学质量与教学改革工程"，将双语课程建设作为拓宽大学生国际视野的一项重要措施。因此，对于高等学校开设双语课程的必要性与重要性已经毋庸置疑。生物化学这门学科具有特殊性，其课程内容建立在英文基础上，因为课程中的实验和理论大多源于国外科学家的研究。此外，生物化学科学研究的最新进展也几乎都以全英文的形式发表在相关的国际专业杂志上。因此，高校开展双语教学有利于学生在系统学习生物化学基础理论的同时提高实际英文应用能力，获取学科前沿知识以及掌握相关的专业英语词汇和表达方式，成为更具有竞争能力的复合型创新人才。对于地方应用型本科高校而言，双语教学显得迫切且具有现实意义。

　　尽管我国各大高校双语教学探索已有若干年，但教材的跟进始终显得滞后。没有合适的教材是目前双语教学面临的困难之一。虽然多数老牌高校喜欢直接采用外语原版教材，但对于地方应用型本科院校来说，单纯使用外语原版教材往往会出现两大问题：一是"水土不服"，外语原版教材中许多针对西方发达国家的案例不适合直接"拿来"；二是"篇幅大、阅读难度大"，学生由此产生畏惧和退缩心理。另外，以英语原版教材作为主、以中文教材为辅的这种"复线型"教材形式存在缺陷，学生往往会依赖中文教材而忽视原版教材的学习。因此，编写小组经过反复讨论与征求学生意见，坚持以外文"原汁原味"为基础，根据引进的经典英文原版教材改编，保证语言纯正；同时考虑双语教学的课时有限，只选取本学科需要重点掌握的内容进行编写，篇幅适中，减轻学生负担。

　　生物化学是多个专业的基础课，不同专业对该课程知识需求的侧重不同。本教材包括三大部分内容，即结构生物化学、新陈代谢和信息大分子。对于食品类专业而言，结构生物化学和新陈代谢是食品专业学生学习的重点，为后续专业课程包括食品化学、食品营养学、食品分析、食品理化检验、食品保藏原理与技术以及功能性食品等奠定基础。对于信息大分子这部分内容,则相对简明扼要,有别于生命科学相关专业,突出应用型人才培养的理念。

本书图文并茂,特别注重双语教学的需要,兼顾学生参差不齐的英语水平,将重要的英文生化专业名词与术语均在文中标出相应的中文。英文段落内容根据其难易程度采取两种方式处理:简单内容用中文进行段落大意归纳,复杂内容采用中文进行段落翻译。本书主要参考的英文教材是 Trudy Mckee 等主编的"Biochemistry：An Introduction(second edition)"。此外,将部分中文教材的内容择其精华编译到本教材中。每个章节之后都附有关键词汇表、思考题等。本书适用于地方应用型本科高校食品科学与工程、食品质量与安全、化学、农学、发酵工程、环境科学等专业进行中英双语教学的学生使用。

由于编者水平有限,时间仓促,书中难免存在不足及疏漏之处,敬请读者批评指正。

编　者

2016 年 8 月

Contents 目录

Introduction 引言 ··· 1

Part I　Structural Biochemistry 结构生物化学

Chapter 1　Carbohydrates 糖类 ·· 7
　1.1　Structure 结构 ··· 8
　1.2　Monosaccharides 单糖 ··· 9
　1.3　Disaccharides 二糖 ··· 13
　1.4　Oligosaccharides and polysaccharides 寡糖及多糖 ······································· 14

Chapter 2　Lipids 脂质 ·· 22
　2.1　Categories of lipids 脂质的类别 ·· 23
　2.2　Biological functions 生物功能 ·· 28
　2.3　Metabolic profiles 代谢概况 ·· 31
　2.4　Nutrition and health 营养与健康 ·· 32

Chapter 3　Proteins and amino acids 蛋白质和氨基酸 ··· 35
　3.1　Proteins 蛋白质 ··· 35
　3.2　Amino acids 氨基酸 ··· 50

Chapter 4　Nucleic acids 核酸 ·· 67
　4.1　Overview of nucleic acid 核酸概述 ··· 67
　4.2　RNA 核糖核酸 ·· 71
　4.3　DNA 脱氧核糖核酸 ·· 79

Chapter 5　Enzymes 酶 ··· 91
　5.1　Enzyme Structure 酶结构 ·· 92
　5.2　Enzyme catalysis 酶催化反应 ··· 93
　5.3　Enzyme kinetics 酶动力学 ··· 95
　5.4　Regulatory enzymes 调节酶 ·· 108

5.5　Enzyme classification 酶的分类 …………………………………………………… 110

5.6　Enzyme application in industry 酶的工业应用 …………………………………… 111

Part II　Metabolism 新陈代谢

Chapter 1　Overview of Metabolism 代谢总论 ……………………………………… 117

1.1　Key biochemicals 重要的生化分子 ………………………………………………… 118

1.2　Catabolism 分解代谢 ………………………………………………………………… 123

1.3　Energy transformations 能量转换 …………………………………………………… 126

1.4　Anabolism 合成代谢 ………………………………………………………………… 127

1.5　Thermodynamics of living organisms 生物热力学 ………………………………… 133

1.6　Regulation and control 调节与控制 ………………………………………………… 133

Chapter 2　Carbohydrate metabolism 糖代谢 ……………………………………… 137

2.1　Glycolysis 糖酵解 …………………………………………………………………… 137

2.2　Gluconeogenesis 糖异生 …………………………………………………………… 157

2.3　Glycogen 糖原 ……………………………………………………………………… 162

2.4　Pentose phosphate pathway 戊糖磷酸途径 ………………………………………… 167

2.5　Citric acid cycle 三羧酸循环 ……………………………………………………… 171

Chapter 3　Lipid metabolism 脂质代谢 ……………………………………………… 180

3.1　Fatty acid synthesis 脂肪酸合成 …………………………………………………… 180

3.2　Fatty acid degradation 脂肪酸降解 ………………………………………………… 195

3.3　Ketone bodies 酮体 ………………………………………………………………… 202

Chapter 4　Nitrogen metabolism 氮代谢 …………………………………………… 205

4.1　Nitrogen fixation 固氮作用 ………………………………………………………… 205

4.2　Amino acid synthesis 氨基酸合成 ………………………………………………… 206

4.3　Nucleotides synthesis 核苷酸合成 ………………………………………………… 210

4.4　Urea cycle 尿素循环 ………………………………………………………………… 216

Chapter 5　Oxidative phosphorylation 氧化磷酸化 ………………………………… 222

5.1　Overview of oxidative phosphorylation 氧化磷酸化概述 ………………………… 222

5.2　Overview of energy transfer by chemiosmosis 通过化学渗透能量转移
概述 ………………………………………………………………………………… 224

5.3　Electron and proton transfer molecules and electron transport chains
电子和质子转移分子和电子传递链 …………………………………………… 225

5.4　ATP synthase (complex V) ATP 合酶(复合体 V) ·················· 238

5.5　Oxidative stress and Inhibitors 氧化应激与抑制剂 ·················· 241

Part III　Informational Macromolecules 信息大分子

Chapter 1　DNA synthesis and repair DNA 合成和修复 ·················· 247
1.1　DNA replication DNA 复制 ·················· 247
1.2　DNA repair DNA 修复 ·················· 256

Chapter 2　RNA biosynthesis and processing RNA 生物合成与加工 ·················· 267
2.1　Transcription 转录 ·················· 267
2.2　Reverse transcription 逆转录 ·················· 273
2.3　mRNA processing 信使 RNA 加工 ·················· 274

Chapter 3　Protein biosynthesis, modifications and degradation 蛋白质生物合成、修饰与降解 ·················· 279
3.1　Translation 翻译 ·················· 279
3.2　Posttranslational modification 翻译后修饰 ·················· 283
3.3　Protein degradation 蛋白质降解 ·················· 285

二维码资源一览表

序号	资源名称	主要知识点	类型	页码
1	生物化学简明双语教程课件	本书三篇共十三章PPT（中英文版）	PPT	版权页
2	蛋白质的结构	1. 蛋白质的一级结构； 2. 蛋白质的二级结构； 3. 蛋白质的超二级结构和结构域； 4. 蛋白质的三级结构； 5. 蛋白质的四级结构。	微课	41
3	染色质：DNA的魔幻空间	1. 染色质； 2. DNA自组装； 3. 核小体结构； 4. 组蛋白结构	微课	81
4	酶动力学	1. 酶的性质； 2. 米-曼氏方程； 3. 酶动力学实验。	动画	95
5	影响酶促反应的因素	1. 酶浓度的影响； 2. 底物浓度的影响； 3. pH的影响； 4. 温度的影响； 5. 激活剂的影响； 6. 抑制剂的影响	微课	103
6	酶的结构与功能的关系	1. 酶的活性部位； 2. 酶的变构部位； 3. 酶原的激活。	微课	110
7	糖酵解	1. 糖酵解的概念； 2. 糖酵解的反应历程； 3. 糖酵解的生物学意义。	微课	137
8	三羧酸循环	1. 三羧酸循环的概念； 2. 三羧酸循环的反应历程； 3. 能量变化； 4. 生物学意义。	微课	171
9	脂肪酸β氧化	1. 脂酸活化； 2. 脂酸的转运； 3. 脂酸的β氧化作用； 4. 能量计算； 5. 脂酸β氧化的调控。	微课	197
10	原核生物的蛋白质生物合成	1. 氨基酸的活化和转运； 2. 肽链合成的起始； 3. 肽链的延长； 4. 肽链的终止和释放。	微课	272

Introduction 引言

Biochemistry, sometimes called **biologicalchemistry**, is the study of chemical processes in living organisms(生物体), including, but not limited to, living matter. Biochemistry governs all living organisms and living processes. By controlling information flow(信息流) through biochemical signalling(生物化学信号) and the flow of chemical energy(化学能量流) through metabolism(代谢), biochemical processes give rise to(导致) the incredible complexity(难以置信的复杂性) of life. Much of biochemistry deals with the structures and functions of cellular components such as proteins(蛋白质), carbohydrates(糖), lipids(脂质), nucleic acids(核酸) and other biomolecules(生物分子) although increasingly processes rather than individual molecules are the main focus. Over the last 40 years biochemistry has become so successful at explaining living processes that now almost all areas of the life sciences(生命科学) from botany(植物) to medicine(医学) are engaged in biochemical research. Today the main focus of pure biochemistry(纯粹生物化学) is in understanding how biological molecules give rise to the processes that occur within living cells which in turn relates greatly to the study and understanding of whole organisms.

生物化学的含义。

生物化学研究对象的复杂性。

生物化学的成就。

当今纯粹生物化学的主要焦点是了解生物分子在活细胞内如何工作,这极大地关系到对整个生物体的研究和理解。

Among the vast number of different biomolecules, many are complex(复合物) and large molecules (called biopolymers 生物大分子,生物聚合物), which are composed of similar repeating subunits(亚单位)(called monomers 单体). Each class of polymeric biomolecule has a different set of subunit types. For example, a protein is a polymer whose subunits are selected from a set of 20 or more amino acids(氨基酸). Biochemistry studies the chemical properties(化学性质) of important biological molecules, like proteins, and in particular the chemistry of enzyme-catalyzed reactions(酶促反应).

生物分子以及生物大分子的含义。

The biochemistry of cell metabolism(新陈代谢) and the endocrine system(内分泌系统) has been extensively(广泛地) described. Other areas of biochemistry include the genetic code (DNA, RNA)(遗传密码), protein synthesis(蛋白质合成), cell membrane transport(细胞跨膜运输), cell membrane transfer and signal transduction(信号转导).

人类对生命物质的认识过程。

It once was generally believed that life and its materials had some essential property(本质属性) or substance distinct from(不同于) any found in non-living matter(非生命物质), and it was thought that only living beings(有机体,生物) could produce the molecules of life. Then, in 1828, Friedrich Wohler published a paper on the synthesis of urea(尿素合成), proving that organic compounds(有机化合物) can be created artificially(人工地).

生物化学的起源及早期发展。

The dawn of biochemistry(生物化学的开端) may have been the discovery of the first enzyme(酶), diastase(淀粉糖化酶)(today called amylase 淀粉酶), in 1833 by Anselme Payen. Eduard Buchner contributed(出版) the first demonstration(论证) of a complex biochemical process(复杂的生化反应) outside of a cell in 1896: alcoholic fermentation(乙醇发酵) in cell extracts(细胞萃取物) of yeast(酵母). Although the term "biochemistry" seems to have been first used in 1882, it is generally accepted that the

"生物化学"学科术语的确立。

formal coinage(正式使用) of biochemistry occurred in 1903 by Carl Neuberg, a German chemist. Previously, this area would have been referred to as(被称为) physiological chemistry(生理化学). Since then, biochemistry has advanced, especially since the mid-20th century, with the development of new techniques such as chromatography(色谱分析法), X-ray diffraction(X射线衍射), dual polarisation interferometry(双偏振干涉法),

促使生物化学迅速发展的诸多相关科学技术。

NMR spectroscopy(核磁共振波谱法), radioisotopic labeling(放射性同位素的标记), electron microscopy(电子显微镜) and molecular dynamics simulations(分子动力学模拟). These techniques allowed for the discovery and detailed analysis of many molecules and metabolic pathways(代谢途径) of the cell, such as glycolysis(糖酵解) and the Krebs cycle (citric acid cycle 三羧酸循环).

Another significant historic event(有历史意义的事件) in biochemistry is the discovery of the gene(基因) and its role in the transfer of information in the cell(在细胞信息传递中所起的作用). This part of biochemistry is often called molecular biology(分子生物学). In the 1950s, James D. Watson, Francis Crick, Rosalind Franklin, and Maurice Wilkins were instrumental(有帮助的) in solving DNA structure(DNA 结构) and suggesting its relationship with genetic transfer of information(遗传信息的传递). In 1958, George Beadle and Edward Tatum received the Nobel Prize for work in fungi(真菌) showing that one gene(基因) produces one enzyme(酶). In 1988, Colin Pitchfork was the first person convicted of murder(谋杀罪名成立) with DNA evidence(DNA 证据), which led to growth of forensic science(司法鉴定). More recently, Andrew Z. Fire and Craig C. Mello received the 2006 Nobel Prize for discovering the role of RNA interference (RNAi)(RNA 干扰), in the silencing(沉默) of gene expression(基因表达).

Today, there are three main types of biochemistry. Plant biochemistry(植物生物化学) involves the study of the biochemistry of autotrophic organisms(自养有机体) such as photosynthesis(光合作用) and other plant specific biochemical processes(特定的生化过程). General biochemistry(普通生物化学) encompasses(包含) both plant and animal biochemistry. Human/medical/medicinal biochemistry(人体/医疗/医药生物化学) focuses on the biochemistry of humans and medical illnesses(疾病).

The four main classes of molecules in biochemistry are carbohydrates, lipids, proteins, and nucleic acids. Many biological molecules are polymers(高分子聚合物); in this terminology(术语), monomers(单分子) are relatively small micromolecules(小分子) that are linked together to create large macromolecules(大分子), which are known as polymers. When monomers are linked together to synthesize(合成) a biological polymer, they undergo a process called dehydration synthesis(脱水缩合).

Part I
Structural Biochemistry

结构生物化学

Chapter 1 Carbohydrates 糖类

Carbohydrate（pronounced /kɑːbəˈhaɪdreɪt/）is an organic compound（有机化合物）with the empirical formula（通式）$C_m(H_2O)_n$（where m could be different from n）; that is, consists only of carbon, hydrogen, and oxygen（仅由 C、H、O 组成）, with a hydrogen:oxygen atom ratio（氢氧原子比）of 2∶1（as in water）. Carbohydrates can be viewed as hydrates（水合物）of carbon, hence their name. Structurally however, it is more accurate to view them as polyhydroxy aldehydes and ketones（多羟基的醛和酮）.

糖的化学组成。

The term is most common in biochemistry, where it is a synonym（同义词）of saccharide（糖）. The carbohydrates（*saccharides*）are divided into four chemical groupings: monosaccharides（单糖）, disaccharides（二糖）, oligosaccharides（寡糖）, and polysaccharides（多糖）. In general, the monosaccharides and disaccharides, which are smaller (lower molecular weight) carbohydrates, are commonly referred to as sugars. The word *saccharide* comes from the Greek word σάκχαρον（*sákkharon*）, meaning "sugar". While the scientific nomenclature（术语）of carbohydrates is complex, the names of the monosaccharides and disaccharides very often end in the suffix-ose（单糖、二糖通常以后缀-ose 结尾）. For example, blood sugar is the monosaccharide glucose, table sugar（蔗糖）is the disaccharide sucrose, and milk sugar is the disaccharide lactose (see illustration（插图）).

糖的分类。

Lactose is a disaccharide found in milk. It consists of a molecule of *D*-galactose and a molecule of *D*-glucose bonded by *beta*-1-4 glycosidic linkage. It has a formula of $C_{12}H_{22}O_{11}$.

乳糖是一种在牛奶中发现的二糖。它由一分子 D-半乳糖和一分子 D-葡萄糖通过 β-1-4 糖苷键组成。

Carbohydrates perform numerous roles(多种角色) in living things. Polysaccharides serve for the storage of energy(贮藏能量)(e.g., starch(淀粉) and glycogen(糖原)), and as structural components(结构成分)(e.g., cellulose in plants(植物中的纤维素) and chitin in arthropods(节肢动物中的几丁质)). The 5 - carbon monosaccharide ribose(核糖) is an important component of coenzymes(辅酶)(e.g., ATP, FAD, and NAD) and the backbone(主链) of the genetic molecule(遗传分子) known as RNA.

The related deoxyribose(脱氧核糖) is a component of DNA. Saccharides and their derivatives(衍生物) include many other important biomolecules that play key roles(重要角色) in the immune system(免疫系统), fertilization(受精), preventing pathogenesis(抗病原体), blood clotting(血液凝固), and development(发育). In food science(食品科学) and in many informal contexts(日常生活中), the term carbohydrate often means any food that is particularly rich in the complex carbohydrate starch (such as cereals, bread, and pasta)(在糖淀粉复合物中含量丰富,例如谷类、面包、面食) or simple carbohydrates, such as sugar (found in candy, jams(果酱), and desserts (甜点)).

1.1 Structure 结构

Formerly(以前) the name "carbohydrate" was used in chemistry for any compound(化合物) with the formula $C_m(H_2O)_n$. Following this definition(由于这个定义), some chemists considered formaldehyde(甲醛) CH_2O to be the simplest carbohydrate, while others claimed that title for glycolaldehyde(羟乙醛). Today the term is generally understood in the biochemistry sense(生物学意义), which excludes(除了) compounds with only one or two carbons.

Natural(自然界) saccharides are generally built of(由……组成) simple carbohydrates called monosaccharides with general formula $(CH_2O)_n$ where n is three or more. A typical(典型的) monosaccharide has the structure H—$(CHOH)_x$—(C═O)—$(CHOH)_y$—H, that is, an aldehyde(醛) or ketone(酮) with many hydroxyl groups(羟基) added, usually one on each carbon atom that is not part of the aldehyde or ketone functional group (官能团). Examples of monosaccharides are glucose(葡萄糖), fructose(果糖), and glyceraldehyde(甘油醛). However, some biological substances commonly called "monosaccharides" do not conform to this formula(不符合通式) (e. g., uronic acids (糖醛酸) and deoxy-sugars(脱氧糖) such as fucose(海藻糖)), and there are many chemicals(化学品) that do conform to this formula but are not considered to be monosaccharides (e. g., formaldehyde CH_2O and inositol(肌糖、纤维糖) $(CH_2O)_6$).

The open-chain form(开链型) of a monosaccharide often coexists with(共存) a closed ring form (闭环型) where the aldehyde/ketone carbonyl(羰基) group carbon (C═O) and hydroxyl group(羟基) (—OH) react forming a hemiacetal(半缩醛) with a new C—O—C bridge.

Monosaccharides can be linked together into what are called polysaccharides (or oligosaccharides) in a large variety of ways(各种各样的方法). Many carbohydrates contain(包含) one or more modified(修饰的) monosaccharide units (单元) that have had one or more groups replaced or removed. For example, deoxyribose(脱氧核糖) a component of DNA, is a modified version of ribose; chitin(几丁质) is composed of repeating units(重复单元) of N-acetylglucosamine(N-乙酰葡糖胺), a nitrogen-containing form of glucose(一个含氮形式的葡萄糖).

1.2 Monosaccharides 单糖

Monosaccharides are the simplest carbohydrates in that they can't be hydrolyzed(水解) to smaller carbohydrates. They are aldehydes or ketones with two or more hydroxyl groups.

The general chemical formula of an unmodified monosaccharide is $(CH_2O)_n$, literally a "carbon hydrate（碳的水合物）". Monosaccharides are important fuel（燃料）molecules as well as building blocks（组成成分）for nucleic acids（核酸）. The smallest monosaccharides, for which $n=3$, are dihydroxyacetone（二羟基丙酮）and D- and L-glyceraldehyde（甘油醛）.

<pre>
    O   H
     \\ //
      C
      |1
  H — C — OH
      |2
 HO — C — H
      |3
  H — C — OH
      |4
  H — C — OH
      |5
     CH₂OH
</pre>

> D-葡萄糖是一种己醛糖，分子式为$(CH_2O)_6$。C^4上连接的—OH位于费歇尔投影式右侧，故为D型糖。

D-glucose is an aldohexose with the formula $(CH_2O)_6$. because C^4—OH is on the right of the Fischer projection, this is a D sugar.

1.2.1　Classification of monosaccharides　单糖的分类

> 单糖分类的三种依据：羰基类型、碳原子数目、手性。

Monosaccharides are classified according to（根据）three different characteristics（特点）: the placement（定位）of its carbonyl group, the number of carbon atoms it contains（包含）, and its chiral handedness（手性）. If the carbonyl group is an aldehyde, the monosaccharide is an aldose（醛糖）; if the carbonyl group is a ketone, the monosaccharide is a ketose（酮糖）. Monosaccharides with three carbon atoms are called trioses（丙糖）, those with four are called tetroses（四糖）, five are called pentoses（戊糖）, six are hexoses（己糖）, and so on. These two systems of classification are often combined（联合）. For example, glucose is an aldohexose（己醛糖）(a six-carbon aldehyde), ribose is an aldopentose（戊醛糖）(a five-carbon aldehyde), and fructose（果糖）is a ketohexose（己酮糖）(a six-carbon ketone).

> 单糖的多个不对称中心导致有多种立体异构体。

Each carbon atom bearing（具有）a hydroxyl group (—OH), with the exception（例外）of the first and last carbons, are asymmetric（不对称的）, making them stereocenters（立体中心）with two possible configurations（构

型)each (*R* or *S*). Because of this asymmetry(不对称), a number of isomers(异构体) may exist for any given monosaccharide formula. The aldohexose(己醛糖) *D*-glucose, for example, has the formula $(CH_2O)_6$, of which all but two of its six carbons atoms are stereogenic(对称的), making *D*-glucose one of $2^4 = 16$ possible stereoisomers(立体异构体). In the case of glyceraldehyde(甘油醛), an aldotriose(丙醛糖), there is one pair of possible stereoisomers, which are enantiomers(对映异构体) and epimers(差向异构体). 1,3-dihydroxyacetone(二羟基丙酮), the ketose corresponding to(相对于) the aldose glyceraldehyde, is a symmetric(对称的) molecule with no stereocenters. The assignment(分配) of *D* or *L* is made according to the orientation(方向,定位) of the asymmetric carbon furthest from the carbonyl group: in a standard Fischer projection if the hydroxyl group is on the right the molecule is a *D* sugar, otherwise(否则) it is an *L* sugar. The "*D*-" and "*L*-" prefixes(前缀) should not be confused with "*d*-" or "*l*-", which indicate(表示) the direction(方向) that the sugar rotates plane polarized light(平面偏振光). This usage(使用) of "*d*-" and "*l*-" is no longer followed in carbohydrate chemistry(不再适用于糖化学).

D 或 *L* 型取决于离羰基最远的非对称碳的取向。

The *α* and *β* anomers(异头物) of glucose. Note the position of the hydroxyl group on the anomeric carbon(异头碳) relative to the CH_2OH group bound to carbon 5: they are either on the opposite sides(异侧) (*α*), or the same side(同侧) (*β*).

葡萄糖的 *α* 和 *β* 型异头物。注意异头碳(即 1 号碳)上的羟基与 5 号碳上连接的 CH_2OH 基团的位置关系:若在异侧为 *α* 型,若在同侧为 *β* 型。

1.2.2 Ring-straight chain isomerism 环-直链异构现象

The aldehyde or ketone group of a straight-chain(直链) monosaccharide will react reversibly(可逆反应) with a hydroxyl group on a different carbon atom to form a hemiacetal(半缩醛) or hemiketal(半缩酮), forming a heterocyclic ring(杂环) with an oxygen bridge between two carbon atoms. Rings with five

直链单糖的醛基或酮基可与其他碳上的羟基发生可逆反应,生成半缩醛或半缩酮,形成氧原子桥接两个碳原子(即 C—O—C)形式的杂环。

and six atoms are called furanose(呋喃糖) and pyranose(吡喃糖) forms, respectively(分别地), and exist in equilibrium(平衡) with the straight-chain form(直链形式).

During the conversion(转换) from straight-chain form to the cyclic(环的) form, the carbon atom containing the carbonyl oxygen(羰基氧原子), called the anomeric carbon(异头碳), becomes a stereogenic center(立体中心) with two possible configurations: the oxygen atom may take a position either above or below the plane of the ring(环平面). The resulting possible pair of stereoisomers are called anomers(由此产生的可能的一对立体异构体被称为异头物). In the α anomer(α异头物), the —OH substituent(取代基) on the anomeric carbon rests(静止) on the opposite side(trans) of the ring from the CH_2OH side branch(分支). The alternative form(另一种形式), in which the CH_2OH substituent and the anomeric hydroxyl are on the same side(cis) of the plane of the ring, is called the β anomer(取代基、异头羟基连在一起的(cis)平面环，称为β异头物). Because the ring and straight-chain forms readily interconvert(互变), both anomers exist in equilibrium(平衡存在).

1.2.3 Use in living organisms 生物体中的用途

Monosaccharides are the major source of fuel for metabolism(代谢燃料的主要来源), being used both as an energy source(能源)(glucose being the most important in nature) and in biosynthesis(生物合成). When monosaccharides are not immediately needed(立即需要) by many cells they are often converted to(转换) more space efficient forms(有效形式), often polysaccharides. In many animals, including humans, this storage form(储存形式) is glycogen(糖原), especially in liver and muscle cells(肝脏和肌肉细胞). In plants(植物中), starch(淀粉) is used for the same purpose.

1.3 Disaccharides 二糖

Two joined monosaccharides are called a disaccharide(二糖) and these are the simplest polysaccharides(多糖). Examples include sucrose(蔗糖) and lactose(乳糖). They are composed of two monosaccharide units bound together by a covalent bond(共价键) known as a glycosidic linkage(糖苷键) formed via(通过) a dehydration reaction(脱水反应), resulting in(导致) the loss of a hydrogen atom(损失一个氢原子) from one monosaccharide and a hydroxyl group(羟基) from the other. The formula of unmodified(未修饰的) disaccharides is $C_{12}H_{22}O_{11}$. Although there are numerous kinds of(多种) disaccharides, a handful of(少量) disaccharides are particularly notable(引人注目).

两个单糖经过脱水反应通过糖苷键连接形成二糖。

Sucrose, also known as table sugar, is a common disaccharide.

Sucrose(蔗糖), pictured above, is the most abundant(最丰富的) disaccharide, and the main form in which carbohydrates are transported in plants(植物中糖类运输的主要形式). It is composed of one D-glucose molecule and one D-fructose molecule. The systematic name for sucrose(蔗糖的系统命名), $O-\alpha-D$-glucopyranosyl$-(1\rightarrow 2)-D$-fructofuranoside, indicates four things:

(1) Its monosaccharides(单糖): glucose(葡萄糖) and fructose.

(2) Their ring types(环类型): glucose is a pyranose(吡喃糖), and fructose is a furanose(呋喃糖).

(3) How they are linked together: the oxygen on carbon number 1 (C^1) of $\alpha-D$-glucose is linked to the C^2 of D-fructose.

蔗糖是最丰富的二糖。

蔗糖由 D-葡萄糖与 D-果糖组成。

蔗糖系统命名包含四种信息。

连接方式:$\alpha-D$-葡萄糖的一号碳(C^1)上的氧连接 D-果糖的二号碳(C^2)。

(4) The -oside suffix(后缀) indicates that the anomeric carbon(异头碳) of both monosaccharides participates in(参与) the glycosidic bond(糖苷键).

Lactose(乳糖), a disaccharide composed of one D-galactose(半乳糖) molecule and one D-glucose molecule, occurs naturally in(天然存在于) mammalian(哺乳动物) milk. The systematic name for lactose is $O\text{-}\beta\text{-}D\text{-}galactopyranosyl\text{-}(1\rightarrow 4)\text{-}D\text{-}glucopyranose$. Other notable disaccharides include maltose(麦芽糖)(two D-glucoses linked $\alpha\text{-}1,4$) and cellulobiose(纤维二糖)(two D-glucoses linked $\beta\text{-}1,4$). disaccharides can be classified into two types. They are reducing and non-reducing disaccahrides. If the functional group(官能团) is present in bonding with another sugar unit, it is called as reducing disaccharide(还原二糖).

常见二糖名称及组成。

1.4 Oligosaccharides and polysaccharides 寡糖及多糖

Oligosaccharides(寡糖) and polysaccharides(多糖) are composed of longer chains(长链) of monosaccharide units bound together by glycosidic bonds(糖苷键). The distinction(区别) between the two is based upon(基于) the number of monosaccharide units present in the chain. Oligosaccharides typically contain between three and ten monosaccharide units, and polysaccharides contain greater than ten monosaccharide units.

寡糖和多糖是由单糖单元通过糖苷键连接形成的长链组成。

直链淀粉是由葡萄糖主要通过 $\alpha(1\rightarrow 4)$ 键连接形成的线性聚合物。它可以由几千个葡萄糖单元组成,是淀粉的两种成分之一,另一种是支链淀粉。

Amylose is a linear polymer of glucose mainly linked with $\alpha(1\rightarrow 4)$ bonds. It can be made of several thousands of glucose units. It is one of the two components of starch, the other being amylopectin.

Oligosaccharides are found as a common form of protein posttranslational modification(翻译后修饰). Polysaccharides represent an important class of biological polymers(生物聚合物). Polysaccharides are polymeric carbohydrate structures, formed of repeating units(重复单元)(either mono- or disaccharides) joined together by glycosidic bonds(糖苷键). These structuresare often linear(线性的), but may contain various degrees of branching(不同程度的分支). Polysaccharides are often quite heterogeneous(成分混杂的), containing slight(轻微的) modifications of the repeating unit. Depending on the structure, these macromolecules can have distinct properties(不同属性) from their monosaccharide building blocks(构成单元). They may be amorphous(无定形的) or even insoluble in water(不溶于水).

When all the monosaccharides in a polysaccharide are the same type the polysaccharide is called a homopolysaccharide(同多糖) or homoglycan, but when more than one type of monosaccharide is present they are called heteropolysaccharides(杂多糖) or heteroglycans.

Examples include storage(储存) polysaccharides such as starch and glycogen(糖原), and structural polysaccharides(结构性多糖) such as cellulose(纤维素) and chitin(几丁质).

Polysaccharides have a general formula(一般通式) of $C_x(H_2O)_y$ where x is usually a large number between 200 and 2500. Considering that the repeating units in the polymer backbone(主干) are often six-carbon monosaccharides, the general formula can also be represented as $(C_6H_{10}O_5)_n$ where $40 \leqslant n \leqslant 3\,000$.

1.4.1 Storage polysaccharides 储存多糖

(1) Starches 淀粉

Starches are glucose polymers(葡萄糖聚合物) in which glucopyranose(吡喃葡萄糖) units are bonded by *alpha*-linkages. It is made up of a mixture(混合) of Amylose(直链淀

粉)(15%~20%) and Amylopectin(支链淀粉)(80%~85%). Amylose consists of a linear chain(直链) of several hundred glucose molecules and amylopectin is a branched molecule made of several thousand glucose units (every chain 24~30 glucose units). Starches are insoluble in water. They can be digested by hydrolysis(水解), catalyzed(催化) by enzymes called amylases (淀粉酶), which can break(打破) the *alpha*-linkages (glycosidic bonds)(糖苷键). Humans and other animals have amylases, so they can digest starches. Potato(马铃薯), rice, wheat(小麦), and maize(玉米) are major sources of starch in the human diet(饮食). The formation of starches are the way that plants store glucose.

淀粉分为直链淀粉和支链淀粉,可通过淀粉酶水解消化。

(2) Glycogen 糖原

Glycogen is a polysaccharide that is found in animals and is composed of a branched chain(支链) of glucose residues(残基). It is stored in liver(肝脏) and muscles(肌肉). It is an energy reserve(储备) for animals. It is the chief(主要) form of carbohydrate stored in animal body. It is insoluble(不溶) in water. It gives red colour with iodine(碘). It also yields(产生) glucose on hydrolysis(水解).

糖原的组成,分布及基本性质。

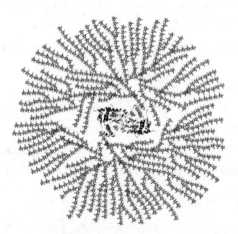

Glycogen.

糖原分子结构图。

1.4.2 Structural polysaccharides 结构多糖

(1) Cellulose 纤维素

Cellulose is used in the cell walls(细胞壁) of plants and

纤维素是植物细胞壁的主要结构组分。

other organisms, and is claimed to be the most abundant organic molecule(有机分子) on earth. It has many uses such as a significant role in the paper and textile industries(纺织行业), and is used as a feedstock(原料) for the production of rayon(人造丝)(via the viscose process 通过粘胶过程), cellulose acetate(醋酸纤维素), celluloid(赛璐珞), and nitrocellulose(硝化纤维). The structural component(结构组分) of plants are formed primarily(主要) from cellulose. Wood(木材) is largely cellulose(纤维素) and lignin(木质素), while paper and cotton are nearly pure(纯的) cellulose. Cellulose is a polymer(聚合物) made with repeated glucose units bonded together by *beta*-linkages. Humans and many other animals lack(缺少) an enzyme to break the *beta*-linkages, so they do not digest cellulose. Certain animals can digest cellulose, because bacteria(细菌) possessing the enzyme are present in their gut(肠道). The classic example is the termite(白蚁). It is insoluble in water. It gives no color with iodine(遇碘不变色). On hydrolysis(水解), it yields glucose. It is the most abundant carbohydrate in nature.

人类及许多动物缺少能断开β链的酶而不能消化纤维素。能消化纤维素的一些动物是因为它们肠道内存在能产生消化酶的细菌。

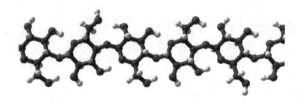

3D structure of cellulose, a *beta*-glucan polysaccharide.

纤维素的三维结构,它是一种β-葡聚糖。

(2) Chitin 几丁质

Chitin is one of many naturally occurring polymers(天然聚合物). It is one of the most abundant natural materials in the world. Over time(随着时间的推移) it is bio-degradable(生物可降解) in the natural environment. Its breakdown may be catalyzed by enzymes called chitinases(几丁质酶), secreted(分泌) by microorganisms(微生物) such as bacteria(细菌) and fungi(真菌), and produced by some plants. Some of these microorganisms have receptors(受体) to simple sugars from the decomposition(分解) of chitin. If chitin is detected, they then

几丁质(甲壳素)是天然聚合物之一,由几丁质酶催化分解。它与纤维素结构类似,都是保护有机体的结构材料。

produce enzymes to digest it by cleaving（切开）the glycosidic bonds（糖苷键）in order to convert it to simple sugars and ammonia（氨）. Chemically, chitin is closely related to chitosan（壳聚糖）(a more water-soluble derivative of chitin 水溶性更强的几丁质衍生物). It is also closely related to cellulose in that it is a long unbranched chain of glucose derivatives（葡萄糖衍生物的无分支长链）. Both materials contribute structure and strength, protecting the organism.

（3）Pectins 果胶

> 果胶是复合多糖，主要存在于陆生植物细胞壁。

Pectins are a family of complex polysaccharides that contain 1,4-linked-*D*-galactosyluronic acid residues（残基）. They are present in most primary cell walls（细胞壁）and in the non-woody parts of terrestrial plants（陆生植物）.

（4）Acidic polysaccharides 酸性多糖

> 酸性多糖含有羧基、磷酸基、硫酸酯基。

Acidic polysaccharides are polysaccharides that contain carboxyl groups（羧基）, phosphate groups（磷酸基）and/or sulfuric ester groups（硫酸酯基）.

（5）Bacterial polysaccharides 细菌多糖

> 细菌多糖的分类、结构及生物作用。

Bacterial（细菌）polysaccharides represent a diverse range（不同的范围）of macromolecules（大分子）that include peptidoglycan（肽聚糖）, lipopolysaccharides（脂多糖）, capsules（荚膜）and exopolysaccharides（胞外多糖）; compounds whose functions range from structural cell-wall components (e. g., peptidoglycan), and important virulence factors（致病因素）(e. g., Poly-N-acetylglucosamine（乙酰氨基葡萄糖）in *S. aureus*（金黄色葡萄球菌）), to permitting the bacterium to survive（存活）in harsh environments（恶劣环境）(e. g., *Pseudomonas aeruginosa*（铜绿假单胞菌）in the human lung). Polysaccharide biosynthesis（生物合成）is a tightly regulated（严格监管）, energy-intensive process（能源密集型过程）and understanding the subtle interplay（巧妙的相互作用）between the regulation（调控）and energy conservation（保护）, polymer modification（聚合物改性）and synthesis（合成）, and the external ecological functions（外部生态功能）is a huge area of research（巨大的研究

> 多糖的生物合成研究有巨大的研究空间、价值和意义。

领域). The potential(潜在的) benefits are enormous(巨大的) and should enable for example the development of novel antibacterial(抗菌) strategies(策略) (e. g., new antibiotics(抗生素) and vaccines(疫苗)) and the commercial exploitation(商业开发) to develop novel applications(新应用).

(6) Bacterial capsular polysaccharides 细菌荚膜多糖

Pathogenic bacteria(病原菌) commonly produce a thick(厚的), mucous-like(黏液样的), layer of polysaccharide. This "capsule"(荚膜) cloaks(掩盖) antigenic proteins(抗原蛋白) on the bacterial surface(细菌表面) that would otherwise provoke(激起) an immune response(免疫反应) and thereby lead to the destruction(破坏) of the bacteria. Capsular polysaccharides are water soluble, commonly acidic(酸性), and have molecular weights(分子量) on the order of 100~1 000 kDa. They are linear(线性) and consist of regularly(有规律的) repeating subunits(亚基) of one to six monosaccharides. There is enormous structural diversity(巨大的结构多样性); nearly two hundred different polysaccharides are produced by *E. Coli*(大肠杆菌) alone. Mixtures(混合物) of capsular polysaccharides, either conjugated(结合的) or native(天然的) are used as vaccines(疫苗).

细菌荚膜多糖性质及结构多样性。

Bacteria and many other microbes(微生物), including fungi(真菌) and algae(藻类), often secrete(分泌) polysaccharides as an evolutionary adaptation(进化适应) to help them adhere(坚持) to surfaces(表面) and to prevent them from drying out(丧失水分). Humans have developed some of these polysaccharides into useful products, including xanthan gum(黄原胶), dextran(右旋糖酐), welan gum(文莱胶), gellan gum(结冷胶) and pullulan(普鲁兰多糖).

多糖对细菌、微生物的生物意义及人类开发的多糖产品。

Most of these polysaccharides exhibit(表现出) interesting and very useful visco-elastic properties(黏弹性特性) when dissolved in water at very low levels(非常低的水平). This gives many foods and various liquid consumer products(各种液体消费品), like lotions(乳液), cleaners(清洁剂) and paints(油漆), for example, a viscous(黏性) appearance when stationary

多糖的物理特性及其应用。

（静止的）, but fluidity（流动性）when the slightest shear（最轻微的剪切）is applied, such as when wiped（擦拭）, poured（倒）or brushed（刷）. This property is referred to as pseudoplasticity（假塑性）, or shear thinning（剪切稀化）. Aqueous solutions（水溶液）of the polysaccharide alone have a curious behavior（奇怪的行为）when stirred（搅拌）. After stopping, the swirl（漩涡）continues due to momentum（动量）, then stops, and then reverses direction（逆转方向）briefly（短暂地）. This recoil（反冲）demonstrates（表明）the elastic effect（弹性效应）of the polysaccharide chains previously streched（之前被拉伸）in solution, returning to their relaxed state.

> 细胞表面多糖在细菌生态学和生理学方面的作用。

Cell-surface（细胞表面）polysaccharides play diverse roles in bacterial ecology and physiology（细菌生态学和生理学）. They serve as a barrier（障碍）between the cell wall and the environment, mediate host-pathogen interactions（调解寄主-病原体相互作用）, and form structural components of biofilms（生物膜）. These polysaccharides are synthesized from nucleotide-activated precursors（核苷酸激活的前体）（called nucleotide sugars 核苷酸糖）and, in most cases, all the enzymes necessary for biosynthesis（生物合成）, assembly（组装）and transport（运输）of the completed polymer are encoded（编码）by genes organized in dedicated clusters（特定簇）within the genome（基因组）of the organism. Lipopolysaccharide（脂多糖）is one of the most important cell-surface polysaccharides, as it plays a key（关键的）structural role in outer membrane integrity（外膜的完整性）, as well as being an important mediator（中介）of host-pathogen（寄主-病原体）interactions.

【Key words】

saccharide 糖	glucose 葡萄糖
aldehyde 醛	galactose 半乳糖
ketone 酮	sucrose 蔗糖
aldose 醛糖	maltose 麦芽糖
ketose 酮糖	lactose 乳糖
monosaccharide 单糖	glycoside 糖苷

disaccharide 二糖	a glycosidic bond 糖苷键
trisaccharide 三糖	hydroxyl group 羟基
oligosaccharide 寡糖	starch 淀粉
polysaccharide 多糖	amylose 直链淀粉
homopolysaccharide 同多糖	amylopectin 支链淀粉
heteropolysaccharide 杂多糖	glycogen 糖原
stereoisomer 立体异构体	cellulose 纤维素
anomer 异头物	chitin 壳多糖，几丁质，甲壳素
conformation 构象	glycoprotein 糖蛋白
isomerization 异构化	proteoglycan 蛋白聚糖
fructose 果糖	lipopolysaccharide 脂多糖

Questions

1. Describe the terms of homopolysaccharides, heteropolysaccharides, glycosidic bond, stereoisomer, and anomer.
2. Why humans can not digest cellulose well?
3. Compare the structural differences between starch and glycogen.

References

[1] McKee Trudy, R McKee James. Biochemistry: An Introduction (Second Edition)[M]. New York: McGraw-Hill Companies, 1999.

[2] Nelson D L, Cox M M. Lehninger Principles of Biochemistry (Third Edition)[M]. Derbyshire: Worth Publishers, 2000.

[3] Ding W, Jia H T. Biochemistry (First Edition)[M]. Beijing: Higher Education Press, 2012.

[4] 郑集，陈钧辉. 普通生物化学(第四版)[M]. 北京：高等教育出版社，2007.

[5] 王艳萍. 生物化学[M]. 北京：中国轻工业出版社，2013.

Chapter 2　Lipids 脂质

脂质的分类。

Lipids are a broad group of naturally occurring molecules which includes fats, waxes(蜡), sterols(植物固醇), fat-soluble vitamins(脂溶性维生素)(such as vitamins A, D, E and K), monoglycerides(单甘酯), diglycerides(双甘酯), phospholipids(磷脂), and others. The main biological functions of lipids include energy storage, as structural(结构) components of cell membranes(细胞膜), and as important signaling molecules(信号分子).

脂质的生物功能。

一些常见脂质的结构。最上面是油酸和胆固醇的结构式，中间是甘油三酯的结构，它是由油酰、硬脂酰和棕榈酰连接甘油主体形成的。底部是常见的磷脂，卵磷脂（磷脂酰胆碱）。

Structures of some common lipids. At the top are oleic acid and cholesterol. The middle structure is a triglyceride composed of oleoyl, stearoyl, and palmitoyl chains attached to a glycerol backbone. At the bottom is the common phospholipid, phosphatidylcholine.

生物脂质依据结构单元的分类。

Lipids may be broadly defined as hydrophobic(疏水性) or amphiphilic(两亲性) small molecules; the amphiphilic nature of some lipids allows them to form structures such as vesicles(囊泡), liposomes(脂质体), or membranes in an aqueous environment(疏水环境). Biological lipids originate entirely or in part from two distinct types of biochemical subunits(子单元) or "building blocks": ketoacyl(酮酯酰) and isoprene(异戊二

烯) groups. Using this approach, lipids may be divided into eight categories(类别): fatty acyls(脂肪酰类), glycerolipids(甘油酯), glycerophospholipids(甘油磷脂), sphingolipids(鞘脂类), saccharolipids and polyketides(聚酮类)(derived from condensation(缩合) of ketoacyl(酮酯酰) subunits); and sterol lipids(甾醇脂) and prenol lipids(戊烯醇脂)(derived from condensation of isoprene subunits).

Although the term lipid is sometimes used as a synonym(同义词) for fats, fats are a subgroup(一群) of lipids called triglycerides(甘油三酯). Lipids also encompass(包含) molecules such as fatty acids and their derivatives (including tri-, di-, and monoglycerides(单甘酯) and phospholipids), as well as other sterol-containing metabolites(代谢物) such as cholesterol(胆固醇). Although humans and other mammals use various biosynthetic pathways to both break down and synthesize(合成) lipids, some essential lipids cannot be made this way and must be obtained from the diet.

脂质和脂肪的区别。

2.1 Categories of lipids 脂质的类别

2.1.1 Fatty acyls 脂肪酰类

Fatty acyls, a generic term for describing fatty acids, their conjugates(轭合物) and derivatives(衍生品), are a diverse(多样化的) group of molecules synthesized(合成) by chain-elongation(延伸) of an acetyl-CoA(乙酰辅酶 A) primer(引物) with malonyl-CoA(丙二酰辅酶 A) or methylmalonyl-CoA groups in a process called fatty acid synthesis. They are made of a hydrocarbon(碳氢化合物) chain that terminates(终止) with a carboxylic(羧基的) acid group; this arrangement confers(赋予) the molecule with a polar, hydrophilic end(亲水末端), and a nonpolar(非极性的), hydrophobic end(疏水末端) that is insoluble in water.

脂肪酰类的定义。

The fatty acid structure is one of the most fundamental

> 脂肪酸是生物脂质最基本的结构,常作为结构更复杂的脂质的组成单元。

categories of biological lipids, and is commonly used as a building block of more structurally complex lipids. The carbon chain, typically between 4 to 24 carbons long, may be saturated (饱和) or unsaturated, and may be attached to functional groups containing oxygen, halogens(卤素), nitrogen(氮) and sulfur(硫). Where a double bond exists, there is the possibility of either a *cis* or *trans* geometric isomerism(几何异构体), which significantly affects the molecule's molecular configuration(分子构型). *Cis*-double bonds cause the fatty acid chain to bend, an effect that is more pronounced the more double bonds there are in a chain. This in turn plays an important role in the structure and function of cell membranes. Most naturally occurring fatty acids are of the *cis* configuration, although the *trans* form does exist in some natural and partially hydrogenated fats and oils.

> 顺式双键导致脂肪酸链弯曲,当链中有更多的双键时,效果会更明显,这相应地又在细胞膜的结构和功能中起着重要的作用。大多数天然存在的脂肪酸是顺式构型,而反式形式存在于一些天然的和部分氢化的脂肪和油中。

Examples of biologically important fatty acids are the eicosanoids(二十烷类), derived primarily from arachidonic acid (花生四烯酸) and eicosapentaenoic(二十碳五烯) acid, which include prostaglandins(前列腺素), leukotrienes(白细胞三烯), and thromboxanes(血栓烷).

2.1.2 Glycerolipids 甘油酯

Glycerolipids are composed mainly of mono-, di- and tri-substituted glycerols(甘油), the most well-known being the fatty acid triesters(三酯) of glycerol (triacylglycerols(甘油三酯)), also known as triglycerides. In these compounds, the three hydroxyl groups of glycerol are each esterified, usually by different fatty acids. Because they function as a food store, these lipids comprise the bulk of storage fat in animal tissues. The hydrolysis of the ester bonds of triacylglycerols and the release of glycerol and fatty acids from adipose tissue is called fat mobilization(脂肪动员).

> 甘油三酯的组成。

> 因为担负着食物储存作用,动物组织中大量的储存脂肪都属于甘油酯。甘油三酯的酯键的水解以及从脂肪组织释放出甘油和脂肪酸被称为脂肪动员。

2.1.3 Glycerophospholipids 甘油磷脂

Glycerophospholipids, also referred to as phospholipids(磷

> 磷脂的分布与重要性。

脂),are ubiquitous(普遍存在) in nature and are key components of the lipid bilayer(脂双层) of cells, as well as being involved in metabolism and cell signaling. Neural tissue(神经组织)(including the brain) contains relatively high amounts of glycerophospholipids, and alterations(改变) in their composition has been implicated in various neurological disorders(神经障碍或紊乱).

Phosphatidylethanolamine(磷脂酰乙醇胺)

Examples of glycerophospholipids found in biological membranes(膜) are phosphatidylcholine(磷脂酰胆碱)(also known as PC, GPCho or lecithin), phosphatidylethanolamine (PE or GPEtn) and phosphatidylserine(磷脂酰丝氨酸)(PS or GPSer). In addition to serving as a primary component(组件) of cellular(细胞) membranes and binding sites for intra - and intercellular proteins, some glycerophospholipids in eukaryotic cells, such as phosphatidylinositols(磷脂酰肌醇) and phosphatidic acids are either precursors of, or are themselves, membrane-derived second messengers.

生物膜磷脂酰胆碱、磷脂酰乙醇胺和磷脂酰丝氨酸除了作为细胞膜的主要成分以及胞内和胞间蛋白的结合位点,在真核细胞中的一些甘油磷脂,如磷脂酰肌醇和磷脂酸,或者作为前体,或者本身作为膜源第二信使。

2.1.4 Sphingolipids 鞘脂

Sphingolipids are a complex family of compounds that share a common structural feature, a sphingoid base backbone that is synthesized(合成) *de novo*(从头) from the amino acid serine (丝氨酸) and a long-chain fatty acyl CoA, then converted(转换) into ceramides(神经酰胺), phosphosphingolipids(磷酸化鞘脂), glycosphingolipids(鞘糖脂) and other compounds. The major sphingoid base(类鞘氨醇碱) of mammals(哺乳动物) is commonly referred to as sphingosine. Ceramides(神经酰胺)

鞘脂类化合物的结构及来源。

(N-acyl-sphingoid bases) are a major subclass of sphingoid base derivatives with an amide-linked fatty acid. The fatty acids are typically saturated or mono-unsaturated with chain lengths from 16 to 26 carbon atoms.

The glycosphingolipids are a diverse family of molecules composed of one or more sugar residues linked via a glycosidic bond(糖苷键) to the sphingoid base(鞘氨醇碱). Examples of these are the simple and complex glycosphingolipids such as cerebrosides(脑苷脂) and gangliosides(神经节苷脂).

Sphingomyelin（鞘磷脂）.

2.1.5　Sterol lipids 固醇脂

固醇脂种类的特点及生物学功能。

Sterol lipids, such as cholesterol(胆固醇) and its derivatives, are an important component of membrane lipids, along with the glycerophospholipids and sphingomyelins. The steroids(类固醇), all derived from the same fused four-ring core structure, have different biological roles as hormones(激素) and signaling molecules(信号分子). The eighteen-carbon (C18) steroids include the estrogen family(雌性激素家族) whereas the C19 steroids comprise the androgens(雄性激素) such as testosterone(睾酮) and androsterone(雄甾酮). The C21 subclass(子类) includes the progestogens(孕激素) as well as the glucocorticoids(糖皮质激素) and mineralocorticoids(盐皮质激素).

所有类固醇都是从同一稠合四环核心结构衍生而来,各有其生物学功能,如作为激素和信号分子。

2.1.6　Saccharolipids 糖脂

糖脂化合物是由脂肪酸直接连接糖骨架形成的。它具有与膜脂双层一致的结构。

Saccharolipids describe compounds in which fatty acids are linked directly to a sugar backbone, forming structures that are compatible with(一致) membrane bilayers. In the saccharolipids, a monosaccharide(单糖) substitutes(取代) for the glycerol backbone present in glycerolipids and glycerophospholipids(甘油磷脂). The most familiar saccharolipids are the acylated glucosamine(酰化氨基葡萄糖) precursors of the Lipid A component of the lipopolysaccharides(脂多糖) in Gram-negative bacteria.

Structure of the saccharo lipid Kdo$_2$-Lipid A.

糖脂 Kdo$_2$-Lipid A 的结构。

Typical lipid A molecules are disaccharides of glucosamine, which are derivatized with as many as seven fatty-acyl chains. The minimal lipopolysaccharide required for growth in *E. col*(大肠杆菌) is Kdo$_2$-Lipid A, a hexa-acylated(六酰化) disaccharide of glucosamine that is glycosylated(糖基化的) with two 3-deoxy(脱氧)-*D*-manno-octulosonic(辛酮糖) acid (Kdo) residues(残基).

2.1.7 Polyketides 聚酮化合物

Polyketides are synthesized by polymerization(聚合) of acetyl(乙酰基) and propionyl(丙酰) subunits by classic enzymes as well as iterative(迭代) and multimodular(多模块化) enzymes that share mechanistic features(机理特征) with the fatty acid synthases. They comprise a large number of secondary metabolites(次级代谢产物) and natural products from animal, plant, bacterial, fungal(真菌) and marine(海洋

聚酮化合物的合成来源及其应用。

的)sources, and have great structural diversity(结构多样). Many polyketides are cyclic molecules(环状分子) whose backbones are often further modified by glycosylation(糖基化), methylation(甲基化), hydroxylation(羟基化), oxidation, and/or other processes. Many commonly used anti-microbial, anti-parasitic(抗寄生虫), and anti-cancer agents are polyketides or polyketide derivatives, such as erythromycins(红霉素), tetracyclines(四环素), avermectins(阿维菌素), and antitumor epothilones(埃坡霉素抗癌药物).

2.2　Biological functions 生物功能

2.2.1　Membranes 细胞膜

甘油磷脂是生物膜,如质膜和细胞器内膜的主要结构组分。动物细胞中,质膜把胞内组分与胞外环境进行了物理分隔。甘油磷脂是两亲分子(同时含疏水性和亲水性结构),其甘油核心结构以酯键连接两个脂肪酸衍生物"尾巴",并以酯键连接一个磷酸"头部"。

Eukaryotic cells(真核细胞) are compartmentalized(划分) into membrane-bound organelles(膜结合细胞器) which carry out different biological functions. The glycerophospholipids(甘油磷脂) are the main structural component of biological membranes, such as the cellular plasma membrane(质膜) and the intracellular membranes of organelles; in animal cells the plasma membrane physically separates(分离) the intracellular components from the extracellular(胞外) environment. The glycerophospholipids are amphipathic molecules(两亲分子) (containing both hydrophobic and hydrophilic regions) that contain a glycerol core linked to two fatty acid-derived "tails" by ester linkages and to one "head" group by a phosphate ester linkage. While glycerophospholipids are the major component of biological membranes, other non-glyceride lipid components such as sphingomyelin(鞘磷脂) and sterols (mainly cholesterol in animal cell membranes) are also found in biological membranes.

Lipid bilayers(磷脂双分子层) have been found to exhibit high levels of birefringence(双折射) which can be used to probe the degree of order(有序度) (or disruption) within the bilayer using techniques such as dual polarisation interferometry(双偏振干涉). A biological membrane is a form of lipid bilayer. The formation of lipid bilayers is an energetically preferred process (能量优先过程) when the glycerophospholipids(甘油磷脂) described above are in an aqueous environment(水相环境).

In an aqueous system, the polar heads of lipids align(排列) towards the polar, aqueous environment, while the hydrophobic tails minimize their contact with water and tend to cluster together, forming a vesicle; depending on the concentration(浓度) of the lipid, this biophysical interaction may result in the formation of micelles(胶束), liposomes, or lipid bilayers. Other aggregations(聚合) are also observed and form part of the polymorphism(多态性) of amphiphile (lipid) behavior(两亲行为). Micelles and bilayers form in the polar medium by a process known as the hydrophobic effect(疏水效应). When dissolving a lipophilic or amphiphilic substance in a polar environment, the polar molecules (i.e., water in an aqueous solution) become more ordered around the dissolved lipophilic substance, since the polar molecules cannot form hydrogen bonds to the lipophilic areas of the amphiphile. So in an aqueous environment, the water molecules form an ordered "clathrate" cage around the dissolved lipophilic molecule.

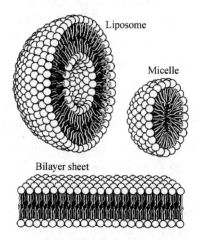

Self-organization of phospholipids: a spherical liposome, a micelle and a lipid bilayer.

2.2.2　Energy storage 能量储存

Triacylglycerols(甘油三酯), stored in adipose tissue, are a major form of energy storage in animals. The adipocyte(脂肪细胞), or fat cell, is designed for continuous synthesis and breakdown of triacylglycerols, with breakdown controlled mainly by the activation of hormone-sensitive(激素敏感) enzyme lipase. The complete oxidation of fatty acids provides high caloric content, about 9 kcal/g, compared with 4 kcal/g for the breakdown of carbohydrates and proteins. Migratory birds(候鸟) that must fly long distances without eating use stored energy of triacylglycerols to fuel their flights.

脂肪酸的完全氧化能提供高达约9千卡/克的热量,相比之下,糖和蛋白质分解提供4千卡/克的热量。候鸟为了进行长距离无摄食的飞行,必须要使用甘油三酯储存的能量帮助飞行。

2.2.3　Signaling 信号传导

In recent years, evidence has emerged showing that lipid signaling is a vital part of the cell signaling. Lipid signaling may occur via activation of G protein-coupled(G蛋白偶联) or nuclear receptors(核受体), and members of several different lipid categories have been identified as signaling molecules and cellular messengers(细胞信使). These include sphingosine-1-phosphate(鞘氨醇-1-磷酸), a sphingolipid derived from ceramide(神经酰胺) that is a potent messenger molecule involved in regulating calcium mobilization(调控钙动员), cell growth, and apoptosis(细胞凋亡); diacylglycerol(甘油二酯)(DAG) and the phosphatidylinositol phosphates(磷脂酰肌醇磷酸盐, PIPs), involved in calcium-mediated activation of protein kinase C(蛋白激酶C); the prostaglandins(前列腺素), which are one type of fatty-acid derived eicosanoid(类花生酸) involved in inflammation(炎症) and immunity(免疫); the steroid hormones such as estrogen(雌激素), testosterone(睾酮) and cortisol(皮质醇), which modulate(调制) a host of functions such as reproduction, metabolism and blood pressure; and the oxysterols(氧化型胆固醇) such as 25-hydroxy-cholesterol(25-羟基胆固醇) that are liver X receptor agonists(受体激动剂).

近年来,有确凿的证据表明脂质信号传导是细胞信号传导的一个重要部分。

2.3 Metabolic profiles 代谢概况

The major dietary lipids for humans and other animals are animal and plant triglycerides, sterols(固醇类), and membrane phospholipids. The process of lipid metabolism synthesizes and degrades the lipid stores and produces the structural and functional lipids characteristic of individual tissues.

2.3.1 Biosynthesis 生物合成

In animals, when there is an oversupply of dietary carbohydrate, the excess carbohydrate is converted to triacylglycerol. This involves the synthesis of fatty acids from acetyl-CoA(乙酰辅酶 A) and the esterification of fatty acids in the production of triacylglycerol, a process called lipogenesis(脂肪生成). Fatty acids are made by fatty acid synthases(脂肪酸合酶) that polymerize and then reduce acetyl-CoA units. The acyl chains in the fatty acids are extended by a cycle of reactions that add the acetyl group, reduce it to an alcohol, dehydrate(脱水) it to an alkene group(烯烃) and then reduce it again to an alkane group(烷烃).

甘油三酯的合成途径。

The enzymes of fatty acid biosynthesis are divided into two groups, in animals and fungi(真菌) all these fatty acid synthase reactions are carried out by a single multifunctional protein(多功能蛋白), while in plant plastids(质粒体) and bacteria separate enzymes perform each step in the pathway(通路). The fatty acids may be subsequently converted to triacylglycerols that are packaged in lipoproteins and secreted(分泌) from the liver.

参与脂肪酸生物合成的酶分为两种类型。

The synthesis of unsaturated fatty acids involves a desaturation reaction(脱氢反应), whereby a double bond is introduced into the fatty acyl chain. For example, in humans, the desaturation of stearic acid(硬脂酸) by stearoyl-CoA desaturase-1 produces oleic acid(油酸). The doubly unsaturated

不饱和脂肪酸生物合成过程。

fatty acid linoleic acid(亚油酸) as well as the triply unsaturated α-linolenic acid(亚麻酸) cannot be synthesized in mammalian tissues(哺乳动物组织), and are therefore essential fatty acids and must be obtained from the diet.

Triacylglycerol synthesis takes place in the endoplasmic reticulum(内质网) by metabolic pathways in which acyl groups in fatty acyl-CoAs(脂酰辅酶A) are transferred to the hydroxyl groups of glycerol-3-phosphate(甘油-3-磷酸) and diacylglycerol(甘油二酯).

2.3.2　Degradation 降解

Beta oxidation is the metabolic process by which fatty acids are broken down in the mitochondria(线粒体) and/or in peroxisomes(过氧化物酶体) to generate acetyl-CoA. For the most part, fatty acids are oxidized by a mechanism that is similar to, but not identical with(等同), a reversal of the process of fatty acid synthesis. That is, two-carbon fragments are removed sequentially from the carboxyl end of the acid after steps of dehydrogenation(脱氢作用), hydration(水化), and oxidation to form a *beta*-keto acid(β-酮酸), which is split by thiolysis(硫解). The acetyl-CoA is then ultimately converted into ATP, CO_2, and H_2O using the citric acid cycle(三羧酸循环) and the electron transport chain(电子传递链).

Hence the Krebs Cycle(三羧酸循环) can start at acetyl-CoA when fat is being broken down for energy if there is little or no glucose available. The energy yield of the complete oxidation of the fatty acid palmitate(棕榈酸酯) is 106 ATP. Unsaturated and odd-chain(奇数链) fatty acids require additional enzymatic steps for degradation.

2.4　Nutrition and health 营养与健康

Most of the lipid found in food is in the form of triacylglycerols(甘油三酯), cholesterol(胆固醇) and phospholipids(磷脂). A minimum amount of dietary fat is necessary to facilitate(促进) absorption of fat-soluble vitamins (A, D, E and K) and carotenoids(类胡萝卜素). Humans and other mammals have a dietary requirement for certain essential fatty acids, such as linoleic acid(亚油酸) (an omega-6 fatty acid) and *alpha*-linolenic acid(α-亚麻酸) (an omega-3 fatty acid) because they cannot be synthesized from simple precursors in the diet.

Both of these fatty acids are 18-carbon polyunsaturated fatty acids(18-碳不饱和脂肪酸) differing in the number and position of the double bonds. Most vegetable oils are rich in linoleic acid (safflower 红花, sunflower 向日葵, and corn oils 玉米油). *Alpha*-linolenic acid is found in the green leaves of plants, and in selected seeds, nuts and legumes (particularly flax 胡麻, rapeseed 胡麻, walnut 核桃 and soy 核桃). Fish oils are particularly rich in the longer-chain omega-3 fatty acid seicosapentaenoic acid(二十碳五烯酸)(EPA) and docosahexaenoic acid(二十二碳六烯酸)(DHA). A large number of studies have shown positive health benefits associated with consumption of omega-3 fatty acids on infant development(婴儿发育), cancer, cardiovascular diseases(心血管疾病), and various mental illnesses(精神疾病), such as depression(抑郁症), attention-deficit hyperactivity disorder(注意力缺陷多动障碍), and dementia(痴呆). In contrast, it is now well-established that consumption of trans fats, such as those present in partially hydrogenated vegetable oils, are a risk factor for cardiovascular disease(心血管疾病).

> 各种18-碳不饱和脂肪酸在食物中的分布。

> 现在已经形成共识,摄入反式脂肪,如在氢化植物油中部分存在的反式脂肪,是心血管疾病的风险因素。

A few studies have suggested that total dietary fat intake is linked to an increased risk of obesity(肥胖症) and diabetes(糖尿病). However, a number of very large studies, including the Women's Health Initiative Dietary Modification Trial, an eight-year study of 49,000 women, the Nurses' Health Study and the Health Professionals Follow-up Study, revealed no such links. None of these studies suggested any connection between percentage of calories from fat and risk of cancer, heart disease or weight gain. The Nutrition Source, a website maintained by

> 脂肪摄入对肥胖和糖尿病等的影响。

许多研究显示脂肪摄入总量并不真正与肥胖和疾病相关。

the Department of Nutrition at the Harvard School of Public Health, summarizes the current evidence on the impact of dietary fat: "Detailed research—much of it done at Harvard—shows that the total amount of fat in the diet isn't really linked with weight or disease."

【Key words】

lipids 脂质	fatty acids 脂肪酸
fats 脂肪	triacylglycerol (triglycerides) 甘油三酯
oil 油	sterol 固醇
wax 蜡	cholesterol 胆固醇
compound lipids 复脂	sphingolipids 鞘脂
glycolipids/saccharolipids 糖脂	sphingosine 鞘氨醇
phospholipids 磷脂	unsaturated fatty acid 不饱和脂肪酸
glycerol 甘油(醇)	essential fatty acids (EFA) 必需脂肪酸
glycerophospholipids 甘油磷脂	amphipathic compound 两亲化合物

Questions

1. Compare the differences between lipids and fats.
2. Describe the terms of amphipathic compounds, triglycerides, and sphingolipids.
3. What are the essential fatty acids?

References

[1] McKee Trudy, R McKee James. Biochemistry: An Introduction (Second Edition)[M]. New York: McGraw-Hill Companies, 1999.

[2] Nelson D L, Cox M M. Lehninger Principles of Biochemistry (Third Edition)[M]. Derbyshire: Worth Publishers, 2000.

[3] Ding W, Jia H T. Biochemistry (First Edition)[M]. Beijing: Higher Education Press, 2012.

[4] 郑集,陈钧辉. 普通生物化学(第四版)[M]. 北京:高等教育出版社,2007.

[5] 王艳萍. 生物化学[M]. 北京:中国轻工业出版社,2013.

Chapter 3　Proteins and amino acids
蛋白质和氨基酸

3.1　Proteins 蛋白质

Proteins are biochemical compounds consisting of one or more polypeptides(多肽) typically folded into a globular or fibrous form(球状或纤维状), facilitating a biological function. A polypeptide is a single linear polymer chain(线性聚合物) of amino acids bonded together by peptide bonds(肽键) between the carboxyl and amino groups of adjacent amino acid residues(相邻氨基酸残基). The sequence(序列) of amino acids in a protein is defined by the sequence of a gene(基因序列), which is encoded in the genetic code(遗传密码). In general, the genetic code specifies 20 standard amino acids(标准氨基酸); however, in certain organisms the genetic code can include selenocysteine(硒代半胱氨酸)—and in certain archaea(古生菌)—pyrrolysine(吡咯赖氨酸). Shortly after or even during synthesis, the residues in a protein are often chemically modified by posttranslational modification(翻译后修饰), which alters the physical and chemical properties, folding(折叠), stability(稳定性), activity(活性), and ultimately, the function of the proteins.

蛋白质组成及结构特点。

蛋白质中氨基酸序列是由遗传密码决定的。

蛋白质上的氨基酸残基经常被化学修饰。

One of the most distinguishing features of polypeptides is their ability to fold into a globular state(球形状态). The extent to(程度) which proteins fold into a defined structure varies widely. Some proteins fold into a highly rigid structure(高刚性结构) with small fluctuations(波动) and are therefore considered to be single structure. Other proteins undergo large rearrangements(重排) from one conformation to another. This conformational change(构象改变) is often associated with a

多肽一种最显著的特征是其折叠形成球状的能力。蛋白质为形成特定结构而折叠的程度变化很大。

signaling event(信号事件). Thus, the structure of a protein serves as a medium(媒介) through which to regulate either the function of a protein or activity(活性) of an enzyme(酶). Not all proteins require a folding process in order to function, as some function in an unfolded state.

蛋白质的结构作为一种媒介,通过它可以调节蛋白质的功能或酶的活性。

Like other biological macromolecules such as polysaccharides(多糖) and nucleic acids(核酸), proteins are essential parts of organisms and participate in virtually every process within cells. Many proteins are enzymes(酶) that catalyze biochemical reactions(催化生化反应) and are vital to metabolism. Proteins also have structural or mechanical functions, such as actin(肌动蛋白) and myosin(肌球蛋白) in muscle and the proteins in the cytoskeleton(细胞骨架), which form a system of scaffolding(支架) that maintains cell shape(维持细胞形状). Other proteins are important in cell signaling(细胞信号转导), immune responses(免疫反应), cell adhesion(细胞黏附), and the cell cycle(细胞周期). Proteins are also necessary in animals' diets, since animals cannot synthesize all the amino acids(氨基酸) they need and must obtain essential amino acids from food. Through the process of digestion(消化), animals break down ingested protein into free amino acids that are then used in metabolism.

蛋白质在细胞中的功能。

蛋白质也是动物膳食所必需的,因为动物不能合成所有必需的氨基酸,而必须从食物中获取。

The central role of proteins as enzymes in living organisms was however not fully appreciated until 1926, when James B. S. showed that the enzyme urease(蛋白酶脲酶) was in fact a protein. The first protein to be sequenced was insulin(胰岛素), by Frederick Sanger(弗雷德里克), who won the Nobel Prize for this achievement in 1958. The first protein structures to be solved were hemoglobin and myoglobin, by Max Perutz and Sir John Cowdery Kendrew, respectively, in 1958. The three-dimensional structures of both proteins were first determined by X-ray diffraction analysis(X-射线衍射分析); Perutz and Kendrew shared the 1962 Nobel Prize in Chemistry for these discoveries. Proteins may be purified from other cellular components using a variety of techniques such as ultracentrifugation(离心), precipitation(沉淀), electrophoresis(电泳), and

蛋白质研究的发展历史。

蛋白质的三维结构是由X-射线衍射分析确定的。

chromatography(色谱法); the advent of genetic engineering(基因工程的到来) has made possible a number of methods to facilitate purification(促进纯化). Methods commonly used to study protein structure and function include immunohisto chemistry(免疫组织化学), site-directed mutagenesis(位点定向诱变), nuclear magnetic resonance and mass spectrometry(核磁共振和质谱).

Distributed computing(分布式计算) is a relatively new tool researchers are using to examine the infamously complex interactions(极其复杂的相互作用) that govern protein folding(控制蛋白质折叠); the statistical analysis techniques(统计分析技术) employed to calculate a protein's probable tertiary structure(四级结构) from its amino acid sequence(氨基酸序列)(primary structure) are well-suited for the distributed computing environment, which has made this otherwise prohibitively expensive(昂贵的) and time consuming problem significantly more manageable(可控的).

3.1.1　Biochemistry 生物化学

Most proteins consist of linear polymers built from series of up to 20 different $L-\alpha-$ amino acids. All proteinogenic amino acids possess common structural features, including an α-carbon to which an amino group(氨基), a carboxyl group(羧基), and a variable side chain(不同侧链) are bonded. Only proline(脯氨酸) differs from this basic structure as it contains an unusual ring to the $N-$ end amine group, which forces the CO—NH amide moiety(酰胺结构) into a fixed conformation. The side chains of the standard amino acids, detailed in the list of standard amino acids, have a great variety of chemical structures and properties; it is the combined effect of all of the amino acid side chains in a protein that ultimately determines its three-dimensional structure and its chemical reactivity.

Chemical structure of the peptide bond (left) and a peptide bond between leucine(亮氨酸) and threonine(苏氨酸)(right).

多肽链中氨基酸通过肽键连接。

The amino acids in a polypeptide chain are linked by peptide bonds. Once linked in the protein chain, an individual amino acid is called a residue(残基), and the linked series of carbon, nitrogen, and oxygen atoms are known as the main chain or protein backbone. The end of the protein with a free carboxyl group(游离羧基) is known as the C-terminus(碳端) or carboxyl terminus, whereas the end with a free amino group(游离氨基) is known as the N－terminus(N端) or amino terminus.

蛋白质的 C 端与 N 端。

蛋白质和多肽的区别。

Protein is generally used to refer to the complete biological molecule in a stable conformation, whereas peptide is generally reserved for a short amino acid oligomers often lacking a stable three-dimensional structure(三维结构). However, the boundary between the two is not well defined and usually lies near 20~30 residues. Polypeptide can refer to any single linear chain of amino acids, usually regardless of length, but often implies an absence of a defined conformation(构象).

3.1.2 Synthesis 合成

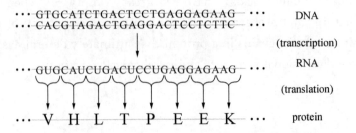

The DNA sequence of a gene encodes the amino acid sequence of a protein.

Proteins are assembled from amino acids using information encoded(编码信息) in genes. Each protein has its own unique amino acid sequence that is specified by the nucleotide sequence(核苷酸序列) of the gene encoding this protein. The genetic code(遗传密码) is a set of three-nucleotide sets called codons(密码子) and each three-nucleotide combination(核苷酸三联体) designates(指定) an amino acid, for example AUG (adenine－uracil－guanine)(腺嘌呤-尿嘧啶-鸟嘌呤) is the code for methionine(蛋氨酸). Because DNA contains four nucleotides, the total number of possible codons is 64; hence, there is some redundancy(冗余) in the genetic code, with some amino acids specified by more than one codon. Genes encoded in DNA are first transcribed into pre-messenger RNA (mRNA) by proteins such as RNA polymerase(聚合酶). Most organisms then process the pre-mRNA (also known as a primary transcript(初级转录物)) using various forms of posttranscriptional modification(转录后修饰) to form the mature(成熟) mRNA, which is then used as a template for protein synthesis by the ribosome(核糖体). In prokaryotes(原核生物) the mRNA may either be used as soon as it is produced, or be bound by a ribosome after having moved away from the nucleoid(拟核). In contrast, eukaryotes(真核生物) make mRNA in the cell nucleus and then translocate it across the nuclear membrane(核膜) into the cytoplasm(胞浆), where protein synthesis then takes place. The rate of protein synthesis is higher in prokaryotes than eukaryotes and can reach up to 20 amino acids per second.

蛋白质是由通过基因编码信息提供的氨基酸组装而成的。

使用各种形式的转录后修饰生成成熟的 mRNA，然后作为核糖体合成蛋白质的模板。

The process of synthesizing a protein from an mRNA template is known as translation. The mRNA is loaded onto the ribosome and is read three nucleotides at a time by matching each codon to its base pairing anticodon(配对的反密码子) located on a transfer RNA(转移 RNA) molecule, which carries the amino acid corresponding to the codon it recognizes. The enzyme aminoacyl tRNA synthetase "charges" the tRNA molecules with the correct amino acids. The growing polypeptide is often

从 mRNA 模板合成蛋白质的过程被称为翻译。

termed the nascent chain(新生链). Proteins are always biosynthesized from N-terminus to C-terminus. The size of a synthesized protein can be measured by the number of amino acids it contains and by its total molecular mass(总相对分子质量), which is normally reported in units of daltons(道尔顿)(synonymous with atomic mass units), or the derivative unit kilodalton (kDa). Yeast proteins(酵母蛋白) are on average 466 amino acids long and 53 kDa in mass. The largest known proteins are the titins(肌联蛋白), a component of the muscle sarcomere(肌肉的肌节), with a molecular mass of almost 3,000 kDa and a total length of almost 27,000 amino acids.

Chemical synthesis 化学合成

Short proteins can also be synthesized chemically by a family of methods known as peptide synthesis(多肽合成), which rely on organic synthesis techniques such as chemical ligation(化学连接) to produce peptides in high yield(高产率).

Chemical synthesis allows for the introduction of non-natural amino acids into polypeptide chains, such as attachment of fluorescent probes to amino acid side chains. These methods are useful in laboratory biochemistry and cell biology(细胞生物学), though generally not for commercial applications. Chemical synthesis is inefficient for polypeptides longer than about 300 amino acids, and the synthesized proteins may not readily assume their native tertiary structure. Most chemical synthesis methods proceed from C-terminus to N-terminus, opposite the biological reaction.

3.1.3　Structure 结构

Further information：Protein structure prediction(蛋白质结构预测)

The crystal structure of the chaperonin(伴侣蛋白) which assists protein folding.

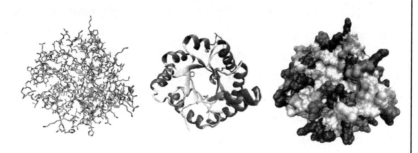

Three possible representations(表示) of the three-dimensional structure of the protein triose phosphate isomerase(磷酸丙糖异构酶). Left: all-atom representation. Middle: Simplified representation illustrating the backbone conformation. Right: Solvent-accessible surface representation.

Most proteins fold into unique 3-dimensional structures(三维结构). The shape into which a protein naturally folds is known as its native conformation(天然构象). Although many proteins can fold unassisted, simply through the chemical properties of their amino acids, others require the aid of molecular chaperones(分子伴侣) to fold into their native states.

Biochemists often refer to four distinct aspects of a protein's structure:

(1) Primary structure(一级结构): the amino acid sequence(氨基酸序列).

(2) Secondary structure(二级结构): regularly repeating local structures stabilized by hydrogen bonds(氢键). The most common examples are the *alpha* helix(α螺旋), *beta* sheet(β折叠) and turns(β转角). Because secondary structures are local(局部的), many regions of different secondary structure can be

绝大多数蛋白质折叠形成独特的三维结构。

微课:蛋白质的结构

二级结构是一种局部结构,所以许多不同的二级结构都可以存在于同一个蛋白质分子中。

present in the same protein molecule.

(3) Tertiary structure(三级结构): the overall shape of a single protein molecule; the spatial relationship of the secondary structures to one another. Tertiary structure is generally stabilized(使稳定) by nonlocal interactions(非局部的相互作用), most commonly the formation of a hydrophobic core(疏水核心), but also through salt bridges(盐桥), hydrogen bonds(氢键), disulfide bonds(二硫键), and even posttranslational modifications(翻译后修饰). The term "tertiary structure" is often used as synonymous with the term *fold*. The tertiary structure is what controls the basic function of the protein.

(4) Quaternary structure(四级结构): the structure formed by several protein molecules (polypeptide chains), usually called *protein subunits* in this context, which function as a single protein complex.

Proteins are not entirely rigid(刚性) molecules. In addition to these levels of structure, proteins may shift between several related structures while they perform their functions. In the context of these functional rearrangements, these tertiary or quaternary structures are usually referred to as "conformations(构象)", and transitions between them are called *conformational changes*. Such changes are often induced by the binding of a substrate molecule(底物分子) to an enzyme's active site, or the physical region of the protein that participates in chemical catalysis. In solution(溶液) proteins also undergo variation in structure through thermal vibration(热振动) and the collision with other molecules.

Proteins can be informally divided into three main classes, which correlate with typical tertiary structures: globular proteins(球状蛋白), fibrous proteins(纤维蛋白), and membrane proteins(膜蛋白). Almost all globular proteins are soluble and many are enzymes. Fibrous proteins are often structural, such as collagen(胶原蛋白), the major component of connective tissue, or keratin(角蛋白), the protein component of hair and nails. Membrane proteins often serve as receptors or provide channels for polar or charged molecules to pass through the cell membrane.

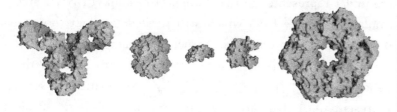

Molecular surface of several proteins showing their comparative sizes. From left to right are: immunoglobulin G(免疫球蛋白)(IgG, an antibody), hemoglobin(血红蛋白), insulin(胰岛素)(a hormone), adenylate kinase(腺苷酸激酶)(an enzyme), and glutamine synthetase(谷氨酰胺合成酶)(an enzyme).

Structure determination 结构测定

Discovering the tertiary structure of a protein, or the quaternary structure of its complexes, can provide important clues about how the protein performs its function. Common experimental methods of structure determination include X-ray crystallography(X-射线晶体学) and NMR spectroscopy(核磁共振光谱), both of which can produce information at atomic resolution(原子分辨率). However, NMR experiments are able to provide information from which a subset of distances between pairs of atoms can be estimated, and the final possible conformations for a protein are determined by solving a distance geometry problem. Dual polarisation interferometry(双偏振干涉) is a quantitative analytical method for measuring the overall protein conformation and conformational changes due to interactions or other stimulus(刺激). Circular dichroism(圆二色谱) is another laboratory technique for determining internal *beta* sheet/*alpha* helical(螺旋) composition of proteins.

Many more gene sequences(基因序列) are known than protein structures. Further, the set of solved structures is biased(偏差) toward proteins that can be easily subjected to the conditions required in X-ray crystallography(X射线晶体学), one of the major structure determination methods. In particular, globular proteins(球状蛋白) are comparatively easy to crystallize in preparation for X-ray crystallography. Membrane proteins, by contrast, are difficult to crystallize and are under represented in the Protein Data Bank(PDB), a freely available resource from which structural data about thousands of proteins can be obtained. Structural genomics(结构基因组学) initiatives(举措) have attempted to remedy these deficiencies(弥补这些不足) by systematically solving representative structures of major fold classes. Protein structure prediction(预测) methods attempt to provide a means of generating a plausible structure(合理的结构) for proteins whose structures have not been experimentally determined.

3.1.4 Cellular functions 细胞功能

Proteins are the chief actors within the cell, said to be carrying out the duties specified by the information encoded in genes(编码基因). With the exception(例外) of certain types of RNA, most other biological molecules are relatively inert elements(惰性元素) upon which proteins act.

Proteins make up half the dry weight of an *Escherichia coli* cell, whereas other macromolecules such as DNA and RNA make up only 3% and 20%, respectively. The set of proteins expressed in a particular cell or cell type is known as its proteome(蛋白质组).

The chief characteristic of proteins that also allows their diverse set of functions is their ability to bind other molecules specifically and tightly(紧密性). The region of the protein responsible for binding another molecule is known as the binding site(结合部位) and is often a depression or "pocket" on the molecular(分子) surface.

This binding ability is mediated by the tertiary structure of the protein, which defines the binding site pocket, and by the chemical properties of the surrounding amino acids' side chains. Protein binding can be extraordinarily tight and specific(特定

的); for example, the ribonuclease(核糖核酸酶) inhibitor(抑制剂) protein binds to human angiogenin(人血管生成素) with a sub-femtomolar dissociation constant(解离常数) (<10⁻¹⁵ M) but does not bind at all to its amphibian homolog onconase(两栖类动物的类似酶) (>1 M).

Proteins can bind to other proteins as well as to small-molecule substrates(底物). When proteins bind specifically to other copies of the same molecule, they can oligomerize(寡聚体化) to form fibrils(纤维); this process occurs often in structural proteins that consist of globular monomers(球状单体) that self-associate to form rigid fibers. Protein-protein interactions also regulate enzymatic activity, control progression through the cell cycle(细胞周期), and allow the assembly of large protein complexes(复合物) that carry out many closely related reactions with a common biological function. Proteins can also bind to, or even be integrated into, cell membranes. The ability of binding partners to induce conformational changes(构象变化) in proteins allows the construction of enormously complex signaling networks(信息网络). Importantly, as interactions between proteins are reversible(可逆的), and depend heavily on the availability of different groups of partner proteins to form aggregates(聚集体) that are capable to carry out discrete(不关联的) sets of function, study of the interactions between specific proteins is a key to understand important aspects of cellular function, and ultimately the properties that distinguish particular cell types.

蛋白质能与蛋白质及小分子底物结合。

蛋白质也能与细胞膜结合或甚至整合到细胞膜中。蛋白质这种结合其他物质以诱导构象变化的能力可构建巨大的复杂信息网络。

研究特定蛋白质间的相互作用是了解细胞功能以及最终区分特定细胞类型的关键。

(1) Enzymes 酶

The best-known role of proteins in the cell is as enzymes, which catalyze chemical reactions. Enzymes are usually highly specific and accelerate only one or a few chemical reactions. Enzymes carry out most of the reactions involved in metabolism(代谢), as well as manipulating(操控) DNA in processes such as DNA replication(复制), DNA repair(修复), and transcription(转录).

蛋白质在细胞中最著名的角色是酶,它催化化学反应。

Some enzymes act on other proteins to add or remove

chemical groups in a process known as posttranslational modification(翻译后修饰). About 4,000 reactions are known to be catalyzed by enzymes. The rate acceleration(加速) conferred by enzymatic catalysis(酶的催化作用) is often enormous(巨大的)—as much as 10^{17}-fold(倍) increase in rate over the uncatalyzed reaction in the case of orotate decarboxylase(乳清酸脱羧酶)(78 million years without the enzyme, 18 milliseconds(毫秒) with the enzyme).

The molecules bound(结合) and acted upon by enzymes are called substrates(底物). Although enzymes can consist of hundreds of amino acids, it is usually only a small fraction of the residues(残基) that come in contact with the substrate, and an even smaller fraction—three to four residues on average—that are directly involved in catalysis. The region of the enzyme that binds the substrate and contains the catalytic residues is known as the active site(活性部位).

(2) Cell signaling and ligand binding 细胞信号传导与配体结合

Many proteins are involved in the process of cell signaling (细胞信号传导) and signal transduction(信号转导). Some proteins, such as insulin(胰岛素), are extracellular proteins(胞外蛋白质) that transmit a signal from the cell in which they were synthesized to other cells in distant tissues(组织). Others are membrane proteins that act as receptors whose main function is to bind a signaling molecule and induce a biochemical response(生化反应) in the cell. Many receptors have a binding site exposed on the cell surface and an effector domain(效应域) within the cell, which may have enzymatic activity(酶的活性) or may undergo a conformational change(构象变化) detected by other proteins within the cell.

Antibodies(抗体) are protein components of an adaptive immune system(免疫系统) whose main function is to bind antigens, or foreign substances in the body, and target them for destruction(消灭).

Ribbon(带状) diagram of a mouse antibody against cholera(霍乱) that binds a carbohydrate antigen.

Antibodies can be secreted(分泌) into the extracellular environment(细胞外环境) or anchored(固定,锚定) in the membranes of specialized B cells known as plasma cells(浆细胞). Whereas enzymes are limited in their binding affinity(亲和力) for their substrates by the necessity of conducting their reaction, antibodies have no such constraints(限制). An antibody's binding affinity to its target is extraordinarily high.

抗体的分泌与结合部位。

Many ligand transport proteins(配体转运蛋白) bind particular small biomolecules(生物分子) and transport them to other locations in the body of a multicellular(多细胞) organism. These proteins must have a high binding affinity when their ligand(配体) is present in high concentrations(高浓度), but must also release the ligand when it is present at low concentrations in the target tissues. The canonical example(典型的例子) of a ligand-binding protein(配体结合蛋白) is hemoglobin(血红蛋白), which transports oxygen(氧) from the lungs(肺) to other organs(器官) and tissues in all vertebrates(脊椎动物) and has close homologs(同系物) in every biological kingdom. Lectins(凝集素) are sugar-binding proteins which are highly specific(高度特异性) for their sugar moieties(糖基). Lectins typically play a role in biological recognition phenomena involving cells and proteins. Receptors(受体) and hormones(激

配体的作用。

配体结合蛋白的典型：血红蛋白。

凝集素的含义及作用。

素)are highly specific binding proteins. Transmembrane proteins(跨膜蛋白)can also serve as ligand transport proteins(配体转运蛋白)that alter the permeability of the cell membrane(细胞膜)to small molecules and ions. The membrane alone has a hydrophobic core(疏水核心)through which polar(极性)or charged(带电)molecules cannot diffuse(扩散). Membrane proteins(膜蛋白)contain internal channels(内部通道)that allow such molecules to enter and exit the cell. Many ion channel proteins(离子通道蛋白)are specialized to select for only a particular ion(特定的离子); for example, potassium and sodium channels(钾和钠离子通道)often discriminate(歧视,选择性通过)for only one of the two ions.

(3) Structural proteins 结构蛋白

Structural proteins(结构蛋白)confer stiffness(僵硬)and rigidity(刚性)to otherwise-fluid biological components. Most structural proteins are fibrous proteins(纤维蛋白); for example, collagen(胶原蛋白)and elastin(弹性蛋白)are critical components of connective tissue(结缔组织)such as cartilage(软骨), and keratin(角蛋白)is found in hard or filamentous(丝状)structures such as hair, nails, feathers, hooves(蹄), and some animal shells. Some globular proteins can also play structural functions, for example, actin and tubulin(肌动蛋白和微管蛋白)are globular(球状)and soluble as monomers(单体), but polymerize(聚合)to form long, stiff fibers that comprise the cytoskeleton(细胞骨架), which allows the cell to maintain its shape and size.

Other proteins that serve structural functions are motor proteins such as myosin(肌球蛋白), kinesin(驱动蛋白), and dynein(动力蛋白), which are capable of generating mechanical forces(机械力). These proteins are crucial for cellular motility(细胞运动)of single celled organisms and the sperm(精子)of many multicellular organisms(多细胞生物)which reproduce sexually. They also generate the forces exerted by contracting muscles(肌肉伸缩).

3.1.5 Methods of study 研究方法

As some of the most commonly studied biological molecules, the activities and structures of proteins are examined both in vitro(体外) and in vivo(体内). In vitro studies of purified proteins in controlled environments are useful for learning how a protein carries out its function: for example, enzyme kinetics studies(酶动力学的研究) explore the chemical mechanism of an enzyme's catalytic activity and its relative affinity for various possible substrate molecules.

结构蛋白的研究方法。

By contrast, in vivo experiments on proteins' activities within cells or even within whole organisms can provide complementary information about where a protein functions and how it is regulated.

体内实验研究的优点。

Protein purification 蛋白质纯化

In order to perform in vitro analysis(体外分析), a protein must be purified away from other cellular components(细胞成分). This process usually begins with cell lysis(裂解), in which a cell's membrane(细胞膜) is disrupted and its internal contents released into a solution known as a crude lysate(粗裂解液). The resulting mixture can be purified using ultracentrifugation(超速离心法), which fractionates(分离) the various cellular components into fractions containing soluble proteins; membrane lipids and proteins; cellular organelles(细胞器), and nucleic acids. Precipitation(沉淀) by a method known as salting out(盐析) can concentrate(浓缩) the proteins from this lysate. Various types of chromatography(色谱法) are then used to isolate the protein or proteins of interest based on properties such as molecular weight, net charge and binding affinity. The level of purification can be monitored using various types of gel electrophoresis if the desired protein's molecular weight and isoelectric point(等电点) are known, by spectroscopy if the protein has distinguishable spectroscopic features(光谱特征), or by enzyme assays if the protein has enzymatic activity. Additionally, proteins can be isolated according their charge using electrofocusing(等电聚焦).

蛋白质纯化过程。

如果已知蛋白质分子量和等电点,或已知其光谱特征,或该蛋白具有酶活性,则蛋白质纯度可以分别通过多种类型的凝胶电泳,光谱法或酶检测法进行监测。

天然蛋白的纯化过程。

利用基因工程在蛋白质上进行化学修饰以简化纯化过程。

For natural proteins, a series of purification steps may be necessary to obtain protein sufficiently pure for laboratory applications. To simplify this process, genetic engineering(基因工程) is often used to add chemical features(化学特征) to proteins that make them easier to purify without affecting their structure or activity. Here, a "tag" consisting of a specific amino acid sequence, often a series of histidine(组氨酸) residues (a "His-tag"), is attached to one terminus(端) of the protein. As a result, when the lysate is passed over a chromatography column (色谱柱) containing nickel(镍), the histidine residues ligate the nickel and attach to the column while the untagged components of the lysate pass unimpeded(无阻碍). A number of different tags have been developed to help researchers purify specific proteins from complex mixtures.

3.2　Amino acids 氨基酸

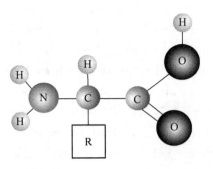

The generic structure of an alpha amino acid in its unionized form.

Amino acids(氨基酸) are molecules(分子) containing an amine group(氨基), a carboxylic acid group(羧基) and a side-chain that varies between different amino acids. The key elements(元素,成分) of an amino acid are carbon, hydrogen (氢), oxygen, and nitrogen(氮). They are particularly important in biochemistry(生物化学), where the term usually refers to *alpha*-amino acids.

An *alpha*-amino acid has the generic formula(通式)

H₂NCHRCOOH, where R is an organic substituent(有机取代基), the amino group is attached to the carbon atom immediately adjacent(邻近的) to the carboxylate group (the α-carbon). Other types of amino acid exist when the amino group is attached to a different carbon atom; for example, in *gamma*-amino acids(γ-氨基酸)(such as *gamma*-amino-butyric acid) the carbon atom to which the amino group attaches is separated from the carboxylate group by two other carbon atoms. The various *alpha*-amino acids differ in which side-chain (R-group) is attached to their *alpha* carbon, and can vary in size from just one hydrogen atom in glycine(甘氨酸) to a large heterocyclic group(杂环基团) in tryptophan(色氨酸).

> 不同α-氨基酸的区别在于连接的α碳原子上的R基。

Amino acids are critical to life, and have many functions in metabolism(代谢). One particularly important function is to serve as the building blocks(组成部分) of proteins(蛋白质), which are linear chains(直链) of amino acids. Amino acids can be linked together in varying sequences(序列) to form a vast variety of proteins. Twenty-two amino acids are naturally incorporated(组成) into polypeptides(多肽) and are called proteinogenic(蛋白氨基酸) or standard amino acids(标准氨基酸). Of these, 20 are encoded(编码) by the universal genetic code(遗传密码).

> 多肽中天然存在22种氨基酸,这些氨基酸被称为蛋白氨基酸或标准氨基酸。其中,20种氨基酸是被通用遗传密码编码的。

Eight standard amino acids are called "essential"(必需的) for humans because they cannot be created from other compounds by the human body, and so must be taken in as food.

> 8种标准氨基酸又被称为人类必需氨基酸,即人体自身不能从其他化合物合成它们,而必须从食物中获得。

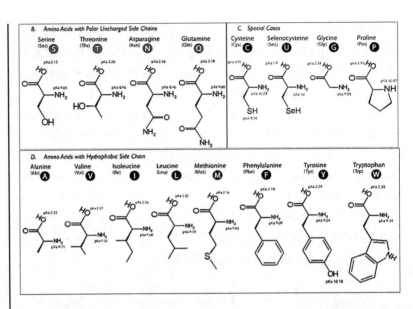

The 21 α-amino acids found in eukaryotes（真核生物）, grouped according to their side-chains' pK_a values and charges carried at physiological pH 7.4.

氨基酸的用途。

Due to their central role in biochemistry, amino acids are important in nutrition（营养）and are commonly used in food technology and industry. In industry, its applications include the production of biodegradable plastics（生物降解塑料）, drugs, and chiral catalysts（手性催化剂）.

3.2.1 General structure 一般结构

在α-氨基酸中,侧链连接在α碳(氨基与羧基连接的碳原子)上。

α-氨基酸中除了甘氨酸, 其他氨基酸中α碳原子都是手性碳原子。在氨基酸中(如赖氨酸,如图所示), 连接在α碳原子上的碳链,其碳原子依次标记为α,β,γ,δ。在一些氨基酸中,依据氨基连接在β或γ碳上,被称为β或γ-氨基酸。

In the structure shown before, R represents a side-chain specific to each amino acid. The carbon atom next to the carboxyl group（羧基）is called the α-carbon and amino acids with a side-chain bonded to this carbon are referred to as *alpha* amino acids. These are the most common form found in nature. In the *alpha* amino acids, the α-carbon is a chiral carbon atom, with the exception of glycine. In amino acids that have a carbon chain attached to the α-carbon (such as lysine（赖氨酸）, shown below) the carbons are labeled（被标记）in order as α, β, γ, δ and so on. In some amino acids, the amine group is attached to the β or γ-carbon, and these are therefore referred to as *beta* or *gamma* amino acids.

$$H_3N^+ \!-\! \overset{\alpha}{\underset{\beta}{\overset{|}{C}}} \!\overset{COO^-}{\underset{|}{\overset{1}{}}} \!-\! H$$

(structure of lysine showing C^1OO^-, $\alpha\,C^2$ with H, $\beta\,CH_2^3$, $\gamma\,CH_2^4$, $\delta\,CH_2^5$, $\varepsilon\,CH_2^6$, NH_3^+)

Lysine with the carbon atoms in the side-chain labeled.

Amino acids are usually classified(分类) by the properties (性质) of their side-chain into four groups. The side-chain can make an amino acid a weak acid(弱酸) or a weak base(弱碱), and a hydrophile(亲水物) if the side-chain is polar(极性的) or a hydrophobe(疏水的) if it is nonpolar(非极性的).

氨基酸依据侧链的性质可被分为四类：酸性氨基酸、碱性氨基酸、极性氨基酸、非极性氨基酸。

The phrase "branched-chain amino acids" or BCAA refers to the amino acids having aliphatic(脂肪族的) side-chains that are non-linear(非直线的); these are leucine(亮氨酸), isoleucine(异亮氨酸), and valine(缬氨酸). Proline(脯氨酸) is the only proteinogenic amino acid whose side-group links to the α‑amino group, and thus, is also the only proteinogenic amino acid containing a secondary amine(仲胺) at this position. In chemical terms, proline is, therefore, an imino acid, since it lacks a primary amino group, although it is still classed as an amino acid in the current biochemical nomenclature(命名法), and may also be called an "N‑alkylated *alpha*-amino acid".

支链氨基酸的定义，种类。

脯氨酸的特殊性。

(1) Isomerism 同分异构

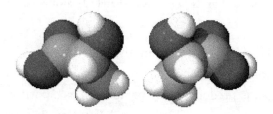

The two optical isomers of alanine, *D* - Alanin and *L* - Alanine.

Of the standard α - amino acids, all but glycine(甘氨酸)

标准α-氨基酸中除了甘氨酸以外均存在两种旋光异构体，即 L 型和 D 型。

can exist in either of two optical isomers(旋光异构体), called L or D amino acids, which are mirror images(镜像) of each other. While L-amino acids represent all of the amino acids found in proteins during translation in the ribosome(核糖体), D-amino acids are found in some proteins produced by enzyme(酶) posttranslational modifications(翻译后修饰) after translation(翻译) and translocation(移动) to the endoplasmic reticulum(内质网), as in exotic sea-dwelling organisms(奇特的海洋生物) such as cone snails(芋螺). They are also abundant components of the peptidoglycan(肽聚糖) cell walls of bacteria(细菌), and D-serine(D-丝氨酸) may act as a neurotransmitter(神经递质) in the brain.

(2) Zwitterions 两性离子

The amine and carboxylic acid functional groups found in amino acids allow them to have amphiprotic(两性的) properties. Carboxylic acid groups (—CO_2H) can be deprotonated(去质子化的) to become negative carboxylates (—CO_2^-), and α-amino groups (—NH_2) can be protonated(质子化) to become positive α-ammonium groups (—NH_3^+). At pH values greater than the pK_a of the carboxylic acid group (mean for the 20 common amino acids is about 2.2), the negative carboxylate ion predominates(占优势).

An amino acid in its (1) unionized and (2) zwitterionic forms.

当 pH 值低于氨基的 pK_a (20 种常见 α 氨基酸的平均值大约为 9.4),氮原子主要为质子化态,形成带正电荷的氨基。

At pH values lower than the pK_a of the α-ammonium group (mean for the 20 common α-amino acids is about 9.4), the nitrogen is predominantly protonated as a positively charged α-ammonium group. Thus, at pH between 2.2 and 9.4, the predominant form adopted by α-amino acids contains a negative carboxylate and a positive α-ammonium group, as shown in structure ② above, so has net zero charge(零净电荷). This molecular state(分子状态) is known as a zwitterion(两性离子), from the German Zwitter meaning hermaphrodite(两性体) or hybrid(混合物). Below pH 2.2, the predominant form(优势形态) will have a neutral(中性的) carboxylic acid group and a positive α-ammonium ion (net charge +1), and above pH 9.4, a negative carboxylate and neutral α-amino group (net charge -1). The fully neutral form (structure ① above) is a very minor species(次要物种) in aqueous solution(水溶液) throughout the pH range (less than 1 part in 10^7(少于 10^7 分之一)). Amino acids also exist as zwitterions in the solid phase(固相), and crystallize(结晶) with salt-like properties unlike typical organic(有机的) acids or amines.

氨基酸具有两性离子性质。

(3) Isoelectric point 等电点

At pH values between the two pK_a values, the zwitterion predominates, but coexists(共存) in dynamic equilibrium(动态平衡) with small amounts of net negative and net positive ions. At the exact midpoint(中点) between the two pK_a values, the trace amount of net negative and trace of net positive ions exactly balance, so that average net charge of all forms present is zero. This pH is known as the isoelectric point(等电点) pI, so $pI = \frac{1}{2}(pK_{a_1} + pK_{a_2})$. The individual amino acids all have slightly different pK_a values, so have different isoelectric points. For amino acids with charged side-chains, the pK_a of the side-chain is involved. Thus for Asp(天冬氨酸), Glu(谷氨酸) with negative side-chains, $pI = \frac{1}{2}(pK_{a_1} + pK_{a_R})$, where pK_{a_R} is the side-chain pK_a. Cysteine(半胱氨酸) also has potentially negative side-chain with $pK_{a_R} = 8.14$, so pI should be calculated as for Asp and Glu, even though the side-chain is not significantly charged at neutral pH. For His(组氨酸), Lys(赖氨酸), and Arg(精氨酸) with positive side-chains, $pI = \frac{1}{2}(pK_{a_R} + pK_{a_2})$. Amino acids have zero mobility(流动性) in electrophoresis(电泳) at their isoelectric point, although this behaviour is more usually exploited(利用) for peptides(多肽类)

氨基酸的等电点 pI 的定义，即

$pI = \frac{1}{2}(pK_{a_1} + pK_{a_2})$

不同氨基酸的 pKa 稍有差别，因此具有不同等电点。

带负电荷侧链的氨基酸：

$pI = \frac{1}{2}(pK_{a_1} + pK_{a_R})$

带正电荷侧链的氨基酸：

$pI = \frac{1}{2}(pK_{a_R} + pK_{a_2})$

在等电点时，氨基酸的特性。

多肽结构特点。

and proteins than single amino acids. Zwitterions have minimum solubility(最小溶解度) at their isolectric point and some amino acids (in particular, with non-polar side-chains) can be isolated(隔离的) by precipitation(沉淀) from water by adjusting the pH to the required isoelectric point.

3.2.2 Occurrence and functions in biochemistry 生物化学中的存在与功能

(1) Standard amino acids 标准氨基酸

Amino acids are the structural units that make up proteins. They join together to form short polymer chains called peptides(肽) or longer chains called either polypeptides(多肽) or proteins. These polymers(聚合物) are linear(线性的) and unbranched(无支链的), with each amino acid within the chain attached to two neighboring amino acids. The process of making proteins is called translation(翻译) and involves the step-by-step addition of amino acids to a growing protein chain by a ribozyme(核糖酶) that is called a ribosome(核糖体). The order in which the amino acids are added is read through the genetic code from an mRNA template(模板), which is a RNA copy of one of the organism's genes.

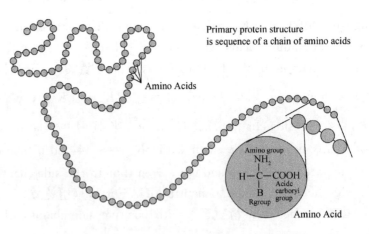

A polypeptide is an unbranched chain of amino acids.

Twenty-two amino acids are naturally incorporated into

polypeptides and are called proteinogenic(蛋白的) or natural amino acids. Of these, 20 are encoded by the universal genetic code. The remaining 2, selenocysteine and pyrrolysine, are incorporated into proteins by unique synthetic mechanisms(合成机制). Selenocysteine(硒代半胱氨酸) is incorporated when the mRNA being translated includes a SECIS element, which causes the UGA codon(密码子) to encode selenocysteine instead of a stop codon(终止密码子). Pyrrolysine(吡咯赖氨酸) is used by some methanogenic archaea(产甲烷古生菌) in enzymes that they use to produce methane(甲烷). It is coded for with the codon UAG, which is normally a stop codon in other organisms.

(2) In human nutrition 营养

When taken up into the human body from the diet, the 22 standard amino acids either are used to synthesize proteins and other biomolecules or are oxidized to urea(尿素) and carbon dioxide as a source of energy. The oxidation pathway starts with the removal(移除) of the amino group by a transaminase(转氨酶), the amino group is then fed into(进入) the urea cycle(尿素循环). The other product of transamidation(转氨基化) is a keto acid(酮酸) that enters the citric acid cycle(柠檬酸循环). Glucogenic amino acids(生糖氨基酸) can also be converted into glucose(葡萄糖), through gluconeogenesis(糖异生).

Of the 22 standard amino acids, 8 are called essential amino acids because the human body cannot synthesize them from other compounds at the level needed for normal growth, so they must be obtained from food. In addition, cysteine(半胱氨酸), tyrosine(酪氨酸), histidine(组氨酸), and arginine(精氨酸) are semiessential amino acids(半必需氨基酸) in children, because the metabolic pathways that synthesize these amino acids are not fully developed. The amounts required also depend on the age and health of the individual, so it is hard to make general statements about the dietary requirement(饮食需求) for some amino acids.

Essential	Nonessential
Isoleucine	Alanine
Leucine	Asparagine
Lysine	Aspartil acid
Methionine	Cysteine*
Phenylalanine	Glutamic acid
Threonine	Glutamine*
Tryptophan	Glycine
Valine	Proline*
	Selenocysteine*
	Serine*
	Tyrosine*
	Arginine*
	Histidine*
	Ornithine*
	Taurine*

(*) Essential only in certain cases.

(3) Reactions 反应

As amino acids have both a primary amine group and a primary carboxyl group, these chemicals can undergo most of the reactions associated with these functional groups. These include nucleophilic addition(亲核加成), amide bond formation (酰胺键的形成) and imine(亚胺) formation for the amine group and esterification（酯化）, amide bond formation and decarboxylation（脱酸）for the carboxylic acid group. The multiple side-chains of amino acids can also undergo chemical reactions. The types of these reactions are determined by the groups on these side-chains and are, therefore, different between the various types of amino acid.

氨基酸同时具有氨基和羧基,能发生氨基和羧基所能进行的反应,如亲核加成,酯化,脱羧等。氨基酸侧链也能进行相应的反应。

3.2.3 Physicochemical properties of amino acids 氨基酸的理化学性质

The 20 amino acids encoded directly by the genetic code(遗传密码) can be divided into several groups based on their properties. Important factors are charge(电荷), hydrophilicity(亲水性) or hydrophobicity(疏水性), size, and functional groups. These properties are important for protein structure and protein-protein interactions. The water-soluble(水溶性) proteins tend to have their hydrophobic residues(残基)(Leu 亮氨酸, Ile 异亮氨酸, Val 缬氨酸, Phe 苯丙氨酸, and Trp 色氨酸) buried in the middle of the protein, whereas hydrophilic(亲水的) side-chains are exposed to the aqueous solvent(水溶剂). The integral membrane proteins tend to have outer rings of exposed hydrophobic amino acids that anchor them into the lipid bilayer(脂质双分子层). In the case part-way between these two extremes(极端例子), some peripheral membrane proteins(外周膜蛋白) have a patch(补丁) of hydrophobic amino acids on their surface that locks onto the membrane. In similar fashion, proteins that have to bind to positively-charged molecules have surfaces rich with negatively charged amino acids like glutamate and aspartate(天冬氨酸), while proteins binding to negatively charged molecules have surfaces rich with positively charged chains like lysine and arginine(精氨酸). There are different hydrophobicity scales(尺度) of amino acid residues.

Some amino acids have special properties such as cysteine, that can form covalent disulfide bonds(二硫共价键) to other cysteine residues, proline that forms a cycle to the polypeptide backbone, and glycine that is more flexible than other amino acids.

Many proteins undergo a range of posttranslational modifications(翻译后修饰), when additional chemical groups are attached to the amino acids in proteins. Some modifications can produce hydrophobic lipoproteins(脂蛋白), or hydrophilic glycoproteins(糖蛋白). These type of modification allow the reversible(可逆的) targeting of a protein to a membrane. For example, the addition and removal of the fatty acid palmitic acid(棕榈酸) to cysteine residues in some signaling proteins(信号蛋白) causes the proteins to attach and then detach(分离) from cell membranes.

> 氨基酸分类的依据：电荷，亲疏水性，尺寸，官能团。

> 水溶性蛋白质倾向于把疏水性残基包埋在内部，而把亲水性残基暴露在水溶剂中。

> 半胱氨酸间可形成二硫键。

Properties of standard amino acids 标准氨基酸的性质

Table of 20 standard amino acid properties.

Amino Acid	Side-chain polarity	Side-chain charge (pH 7.4)	Hydropathy index(亲水指数)	Absorbance λ_{max} (nm)
Alanine	nonpolar	neutral	1.8	
Arginine	polar	positive	−4.5	
Asparagine	polar	neutral	−3.5	
Aspartic acid	polar	negative	−3.5	
Cysteine	polar	neutral	2.5	250
Glutamic acid	polar	negative	−3.5	
Glutamine	polar	neutral	−3.5	
Glycine	nonpolar	neutral	−0.4	
Histidine	polar	positive(10%) neutral(90%)	−3.2	211
Isoleucine	nonpolar	neutral	4.5	
Leucine	nonpolar	neutral	3.8	
Lysine	polar	positive	−3.9	
Methionine	nonpolar	neutral	1.9	
Phenylalanine	nonpolar	neutral	2.8	257,206,188
Proline	nonpolar	neutral	−1.6	
Serine	polar	neutral	−0.8	
Threonine	polar	neutral	−0.7	
Tryptophan	nonpolar	neutral	−0.9	280,219
Tyrosine	polar	neutral	−1.3	274,222,193
Valine	nonpolar	neutral	4.2	

Table of gene expression and biochemistry for 22 standard amino acids.

Amino Acid	Short	Abbrev.	Codon(s)	Occurrence in human proteins (%)	Essential in humans
Alanine	A	Ala	GCU, GCC, GCA, GCG	7.8	—
Cysteine	C	Cys	UGU, UGC	1.9	Conditionally
Aspartic acid	D	Asp	GAU, GAC	5.3	—
Glutamic acid	E	Glu	GAA, GAG	6.3	Conditionally
Phenylalanine	F	Phe	UUU, UUC	3.9	Yes

(续表)

Amino Acid	Short	Abbrev.	Codon(s)	Occurrence in human proteins (%)	Essential in humans
Glycine	G	Gly	GGU, GGC, GGA, GGG	7.2	—
Histidine	H	His	CAU, CAC	2.3	Conditionally
Isoleucine	I	Ile	AUU, AUC, AUA	5.3	Yes
Lysine	K	Lys	AAA, AAG	5.9	Yes
Leucine	L	Leu	UUA, UUG, CUU, CUC, CUA, CUG	9.1	Yes
Methionine	M	Met	AUG	2.3	Yes
Asparagine	N	Asn	AAU, AAC	4.3	—
Pyrrolysine	O	Pyl	UAG*		—
Proline	P	Pro	CCU, CCC, CCA, CCG	5.2	—
Glutamine	Q	Gln	CAA, CAG	4.2	—
Arginine	R	Arg	CGU, CGC, CGA, CGG, AGA, AGG	5.1	Conditionally
Serine	S	Ser	UCU, UCC, UCA, UCG, AGU, AGC	6.8	—
Threonine	T	Thr	ACU, ACC, ACA, ACG	5.9	Yes
Selenocysteine	U	Sec	UGA		—
Valine	V	Val	GUU, GUC, GUA, GUG	6.6	Yes
Tryptophan	W	Trp	UGG	1.4	Yes
Tyrosine	Y	Tyr	UAU, UAC	3.2	Conditionally
Stop codon	—	Term	UAA, UAG, UGA	—	—

Remarks of standard amino acids 标准氨基酸的特点

Amino Acid	Remarks
Alanine	Very abundant, very versatile. More stiff than glycine, but small enough to pose only small steric limits(空间限制) for the protein conformation(构象). It behaves fairly neutrally, and can be located in both hydrophilic regions on the protein outside and the hydrophobic areas inside.
Asparagine or aspartic acid	A placeholder when either amino acid may occupy a position.

(续表)

Amino Acid	Remarks
Cysteine	The sulfur(硫) atom bonds readily to heavy metal ions. Under oxidizing conditions, two cysteines can join together in a disulfide bond(二硫键) to form the amino acid cystine. When cystines are part of a protein, insulin(胰岛素) for example, the tertiary structure(三级结构) is stabilized, which makes the protein more resistant to denaturation(变性); therefore, disulfide bonds are common in proteins that have to function in harsh environments including digestive enzymes(消化酶) (e.g., pepsin and chymotrypsin(胃蛋白酶和胰凝乳蛋白酶)) and structural proteins (e.g., keratin(角蛋白)). Disulfides are also found in peptides too small to hold a stable shape on their own (e.g. insulin).
Aspartic acid	Behaves similarly to glutamic acid. Carries a hydrophilic acidic group with strong negative charge. Usually is located on the outer surface of the protein, making it water-soluble. Binds to positively-charged molecules and ions, often used in enzymes to fix the metal ion. When located inside of the protein, aspartate and glutamate are usually paired with arginine and lysine.
Glutamic acid	Behaves similar to aspartic acid. Has longer, slightly more flexible side chain.
Phenylalanine	Essential for humans. Phenylalanine, tyrosine, and tryptophan contain large rigid aromatic group(芳香基) on the side-chain. These are the biggest amino acids. Like isoleucine, leucine and valine, these are hydrophobic and tend to orient towards the interior(内部) of the folded protein molecule. Phenylalanine can be converted into Tyrosine.
Glycine	Because of the two hydrogen atoms at the α-carbon, glycine is not optically active. It is the smallest amino acid, rotates easily, adds flexibility to the protein chain. It is able to fit into the tightest spaces, e.g., the triple helix of collagen(胶原蛋白). As too much flexibility is usually not desired, as a structural component it is less common than alanine.
Histidine	In even slightly acidic conditions protonation(质子化) of the nitrogen occurs, changing the properties of histidine and the polypeptide as a whole. It is used by many proteins as a regulatory mechanism, changing the conformation and behavior of the polypeptide in acidic regions such as the late endosome(晚期胞内体) or lysosome(溶酶体), enforcing conformation change in enzymes. However only a few histidines are needed for this, so it is comparatively scarce(稀缺).
Isoleucine	Essential for humans. Isoleucine, leucine and valine have large aliphatic hydrophobic side chains. Their molecules are rigid, and their mutual hydrophobic interactions are important for the correct folding of proteins, as these chains tend to be located inside of the protein molecule.
Leucine or isoleucine	A placeholder when either amino acid may occupy a position.

(续表)

Amino Acid	Remarks
Lysine	Essential for humans. Behaves similarly to arginine. Contains a long flexible side-chain with a positively-charged end. The flexibility of the chain makes lysine and arginine suitable for binding to molecules with many negative charges on their surfaces. E. g. , DNA-binding proteins have their active regions rich with arginine and lysine. The strong charge makes these two amino acids prone to be located on the outer hydrophilic surfaces of the proteins; when they are found inside, they are usually paired with a corresponding negatively-charged amino acid, e. g. , aspartate or glutamate.
Leucine	Essential for humans. Behaves similar to isoleucine and valine. See isoleucine.
Methionine	Essential for humans. Always the first amino acid to be incorporated into a protein; sometimes removed after translation. Like cysteine, contains sulfur, but with a methyl group instead of hydrogen. This methyl group can be activated, and is used in many reactions where a new carbon atom is being added to another molecule.
Asparagine	Similar to aspartic acid. Asn contains an amide group where Asp has a carboxyl.
Pyrrolysine	Similar to lysine, with a pyrroline ring attached.
Proline	Contains an unusual ring to the N-end amine group, which forces the CO—NH amide sequence into a fixed conformation. Can disrupt protein folding structures like α helix or β sheet, forcing the desired kink in the protein chain. Common in collagen, where it often undergoes a posttranslational modification to hydroxyproline.
Glutamine	Similar to glutamic acid. Gln contains an amide group where Glu has a carboxyl. Used in proteins and as a storage for ammonia. The most abundant amino acid in the body.
Arginine	Functionally similar to lysine.
Serine	Serine and threonine have a short group ended with a hydroxyl group. Its hydrogen is easy to remove, so serine and threonine often act as hydrogen donors in enzymes. Both are very hydrophilic, therefore the outer regions of soluble proteins tend to be rich with them.
Threonine	Essential for humans. Behaves similarly to serine.
Selenocysteine	Selenated form of cysteine, which replaces sulfur.
Valine	Essential for humans. Behaves similarly to isoleucine and leucine. See isoleucine.
Tryptophan	Essential for humans. Behaves similarly to phenylalanine and tyrosine (see phenylalanine). Precursor(前体) of serotonin. Naturally fluorescent(荧光).
Unknown	Placeholder when the amino acid is unknown or unimportant.

(续表)

Amino Acid	Remarks
Tyrosine	Behaves similarly to phenylalanine (precursor to Tyrosine) and tryptophan (see phenylalanine). Precursor of melanin, epinephrine, and thyroid hormones. Naturally fluorescent, although fluorescence is usually quenched by energy transfer to tryptophans.
Glutamic acid or glutamine	A placeholder when either amino acid may occupy a position.

The structures and abbreviations of 20 standard amino acids.

【Key words】

amino acids 氨基酸	peptide unit 肽单元
peptide bonds 肽键	α-helix α 螺旋
carboxyl group 羧基	β-sheet β 折叠
amino group 氨基	β-turn β-转角
hydrophobic 疏水的	isoelectric point 等电点
hydrophilic 亲水的	zwitterions 两性离子
basic amino acids 碱性氨基酸	hemoglobin (Hb) 血红蛋白
acidic amino acids 酸性氨基酸	molecular chaperones 分子伴侣
primary structure 一级结构	proteomics 蛋白质组学
secondary structure 二级结构	isolation 分离
tertiary structure 三级结构	purification 纯化
quaternary structure 四级结构	essential amino acids 必需氨基酸

Questions

1. The structural level that functional proteins have at least is ( ).
   A. primary structure
   B. secondary structure
   C. tertiary structure
   D. quaternary structure

2. Which pH solution of the following doserum albumins (pI=4.7) have positive charges in ( )
   A. pH4.0
   B. pH5.0
   C. pH6.0
   D. pH7.0

3. The secondary structure of protein includes _____ 、β-sheet、_____ 、and random coil.

4. The chemical bond to maintain primary structure of proteins is ( )
   A. peptide bond
   B. hydrogen bond
   C. hydrophobic bond
   D. salt bond

5. What are the essential amino acids?

References

[1] McKee Trudy, R McKee James. Biochemistry: An Introduction (Second Edition)[M]. New York: McGraw-Hill Companies, 1999.

[2] Nelson D L, Cox M M. Lehninger Principles of Biochemistry (Third Edition)[M]. Derbyshire: Worth Publishers, 2000.

[3] Ding W, Jia H T. Biochemistry (First Edition)[M]. Beijing: Higher Education Press, 2012.

[4] 郑集, 陈钧辉. 普通生物化学(第四版)[M]. 北京: 高等教育出版社, 2007.

[5] 王艳萍. 生物化学[M]. 北京: 中国轻工业出版社, 2013.

Chapter 4　Nucleic acid 核酸

4.1　Overview of nucleic acid 核酸概述

Nucleic acids are biological molecules essential for life, and include DNA (deoxyribonucleic acid 脱氧核糖核酸) and RNA (ribonucleic acid 核糖核酸). Together with proteins, nucleic acids make up the most important macromolecules(大分子); each is found in abundance(大量的) in all living things, where they function in encoding(编码), transmitting(传递) and expressing genetic information(遗传信息). Nucleic acids were first discovered by Friedrich Miescher in 1869. Experimental studies of nucleic acids constitute(组成) a major part of modern biological and medical research, and form a foundation(基础) for genome(基因组) and forensic science(法医学), as well as the biotechnology(生物技术) and pharmaceutical industries(制药工业).

核酸分为 DNA 和 RNA。

核酸研究的意义。

4.1.1　Occurrence and nomenclature 出现和命名

The term *nucleic acid* is the overall name for DNA and RNA, members of a family of biopolymers(生物高分子), and is synonymous(同义的) with polynucleotide(多核苷酸). Nucleic acids were named for their initial(初始的) discovery within the cell nucleus(细胞核), and for the presence of phosphate groups(磷酸基)(related to phosphoric acid 磷酸). Although first discovered within the nucleus(细胞核) of eukaryotic cells(真核细胞), nucleic acids are now known to be found in all life forms, including within bacteria(细菌), archaea(古生菌), mitochondria(线粒体), chloroplasts(叶绿体), viruses(病毒) and viroids(类病毒). All living cells and organelles(细胞器) contain both DNA and RNA, while viruses(病毒) contain

核酸的含义及分布。

either DNA or RNA, but usually not both. The basic component of biological nucleic acids is the nucleotide(核苷酸), each of which contains a pentose sugar(戊糖)(ribose 核糖 or deoxyribose 脱氧核糖), a phosphate group(磷酸基团), and a nucleobase(碱基). Nucleic acids are also generated within the laboratory, through the use of enzymes(酶)(DNA and RNA polymerases 聚合酶) and by solid-phase chemical synthesis(固相合成合成). The chemical methods also enable the generation of altered(变异的) nucleic acids(核酸) that are not found in nature, for example, peptide nucleic acids(肽核酸).

4.1.2 Molecular composition and size 分子构成与大小

Nucleic acids(核酸) can vary in size, but are generally very large molecules. Indeed, DNA molecules are probably the largest individual molecules known. Well-studied biological nucleic acid molecules range in size from 21 nucleotides(核苷酸)(small interfering RNA) to large chromosomes(染色体)(human chromosome is a single molecule that contains 247 million base pairs(碱基对)).

In most cases, naturally occurring DNA molecules are double-stranded(双螺旋) and RNA molecules are single-stranded(单链). There are numerous exceptions(例外), however some viruses(病毒) have genomes(基因组) made of double-stranded(双链的) RNA and other viruses(病毒) have single-stranded(单链的) DNA genomes, and in some circumstances, nucleic acid(核酸) structures with three or four strands(链) can form.

Nucleic acids are linear polymers of nucleotides. Each nucleotide consist of three components: a purine(嘌呤) or pyrimidine(嘧啶) nucleobase(碱基)(sometimes termed nitrogenous base or simply base), a pentose sugar(戊糖), and a phosphate group(磷酸基团). The substructure(子结构) consisting of a nucleobase(碱基) plus sugar is termed a nucleoside(核苷). Nucleic acid types differ in the structure of the sugar in their nucleotides(核苷酸)—DNA contains 2′-

deoxyribose(脱氧核糖), while RNA contains ribose(核糖) (where the only difference is the presence of a hydroxyl group 羟基). Also, the nucleobases found in the two nucleic acid types are different: adenine(腺嘌呤), cytosine(胞嘧啶), and guanine(鸟嘌呤) are found in both RNA and DNA, while thymine(胸腺嘧啶) occurs in DNA and uracil(尿嘧啶) occurs in RNA.

> DNA 与 RNA 组分区别：DNA 中是脱氧核糖，RNA 是核糖；DNA 中有胸腺嘧啶，RNA 中有尿嘧啶。

The sugars and phosphates in nucleic acids are connected to each other in an alternating(交替的) chain (sugar-phosphate backbone) through phosphodiester linkages(磷酸二酯键). In conventional(常规的) nomenclature(命名法), the carbons to which the phosphate groups attach are the 3′- end and the 5′- end carbons of the sugar. This gives nucleic acids directionality(定向), and the ends of nucleic acid molecules are referred to as 5′- end and 3′- end. The nucleobases are joined to the sugars via an N - glycosidic(糖苷的) linkage involving a nucleobase ring nitrogen(N - 1 for pyrimidines(嘧啶) and N - 9 for purines(嘌呤)) and the 1′ carbon of the pentose sugar ring.

> 核酸命名中对核酸结构中头尾的规定。

Non-standard(非标准的) nucleosides(核苷) are also found in both RNA and DNA and usually arise from modification(修饰) of the standard nucleosides within the DNA molecule or the primary (initial) RNA transcript(RNA 转录). Transfer RNA (转运 RNA，tRNA) molecules contain a particularly large number of modified nucleosides(核苷).

> 非标准核苷由标准核苷被修饰产生。

4.1.3　Topology 拓扑结构

Double-stranded nucleic acids are made up of complementary sequences(互补序列), in which extensive Watson-Crick base pairing(碱基配对) results in the formation of a highly repeated and quite uniform(高度重复和非常规则的) double-helical(双螺旋) three-dimensional structure(三维结构). In contrast(与此相反), single-stranded RNA and DNA molecules are not constrained to(局限于) a regular double helix, and can adopt highly complex three-dimensional structures that are based on short stretches(小段) of intramolecular(分子内的) base-paired

> 单链 RNA 和 DNA 分子不再局限于形成规则的双螺旋，而能适应基于分子内的配对碱基序列片段，包括沃森-克里克和非标准碱基对形成的高度复杂的三维结构，以及适应广泛而复杂的三级相互作用。

sequences that include both Watson-Crick and noncanonical(非标准的) base pairs, as well as a wide range of complex tertiary(三级的) interactions(相互作用).

Nucleic acid molecules are usually unbranched(无支链的), and may occur as linear(线型的) and circular(圆形的) molecules. For example, bacterial(细菌的) chromosomes(染色体), plasmids(质粒,质体), mitochondrial(线粒体的) DNA and chloroplast(叶绿体的) DNA are usually circular double-stranded DNA molecules, while chromosomes(染色体) of the eukaryotic(真核生物的) nucleus(细胞核) are usually linear double-stranded DNA molecules. Most RNA molecules are linear, single-stranded molecules, but both circular and branched molecules can result from RNA splicing reactions(剪接反应).

4.1.4 Nucleic acid sequences 核酸序列

One DNA or RNA molecule differs from another primarily in the sequence of nucleotides(核苷酸). Nucleotide sequences are of great importance in biology since they carry the ultimate(最终的) instructions(指令) that encode(编码) all biological molecules, molecular assemblies(分子组装), subcellular(亚细胞的) and cellular structures, organs(器官) and organisms(有机体), and directly enable cognition(认知), memory and behavior. Enormous efforts have gone into the development of experimental methods to determine the nucleotide sequence of biological DNA and RNA molecules, and today hundreds of millions of nucleotides are sequenced daily at genome(基因组) centers and smaller laboratories worldwide.

4.1.5 Types of nucleic acids 核酸类型

(1) Deoxyribonucleic acid 脱氧核糖核酸

Deoxyribonucleic acid(脱氧核糖核酸, DNA) is a nucleic acid that contains the genetic instructions(指令) used in the development and functioning of all known living organisms. The main role of DNA molecules is the long-term storage of information and DNA is often compared to a set of blueprints

（蓝图），since it contains the instructions needed to construct（构造）other components（组件）of cells, such as proteins and RNA molecules. The DNA segments（片段）that carry this genetic information（遗传信息）are called genes（基因），but other DNA sequences have structural purposes, or are involved in regulating the use of this genetic information.

(2) Ribonucleic acid 核糖核酸

Ribonucleic acid（RNA）functions in converting genetic information from genes into the amino acid sequences of proteins. The three universal（普遍的）types of RNA include transfer RNA（转运 RNA，tRNA），messenger RNA（信使 RNA，mRNA），and ribosomal RNA（核糖体 RNA，rRNA）. Messenger RNA acts to carry genetic sequence information between DNA and ribosomes, directing protein synthesis. Ribosomal RNA is a major component of the ribosome, and catalyzes peptide bond formation（肽键生成）. Transfer RNA serves as the carrier molecule for amino acids to be used in protein synthesis, and is responsible for decoding（解码）the mRNA. In addition, many other classes of RNA are now known.

> RNA 分子负责将基因信息转换成蛋白质的氨基酸序列。
> RNA 分子的类型及其重要生物学意义。

(3) Artificial nucleic acid analogs 人工类核酸物

Artificial nucleic acid analogs（类似物）have been designed and synthesized by chemists, and include peptide nucleic acid（肽核酸），morpholino-（吗啉）and locked nucleic acid（锁核酸），as well as glycol（乙二醇）nucleic acid and threose（苏阿糖）nucleic acid. Each of these is distinguished from naturally-occurring DNA or RNA by changes to the backbone of the molecule.

4.2　RNA 核糖核酸

Ribonucleic acid（核糖核酸），or RNA, is one of the three major macromolecules (along with DNA and proteins) that are essential for all known forms of life. Like DNA, RNA is made

> RNA 定义，组成及生物学意义。

up of a long chain of components called nucleotides(核苷酸). Each nucleotide consists of a nucleobase(碱基)(sometimes called a nitrogenous base), a ribose sugar(戊糖), and a phosphate group(磷酸基团). The sequence of nucleotides allows RNA to encode(编码) genetic information. All cellular organisms use messenger RNA(mRNA) to carry the genetic information that directs the synthesis of proteins. In addition, some viruses(病毒) use RNA instead of DNA as their genetic material(遗传物质); perhaps a reflection(反映) of the suggested key role of RNA in the evolutionary(进化的) history of life on Earth.

RNA 在蛋白质合成中的作用。

Like proteins, some RNA molecules play an active role in cells by catalyzing(催化) biological reactions, controlling gene expression(基因表达), or sensing(传感) and communicating responses(交流响应) to cellular signals. One of these active processes is protein synthesis, a universal function whereby(通过) mRNA molecules direct the assembly(组装) of proteins on ribosomes(核糖体). This process uses transfer RNA(tRNA) molecules to deliver amino acids to the ribosome, where ribosomal RNA(rRNA) links amino acids together to form proteins.

RNA 的化学结构及其与 DNA 的不同。

The chemical structure of RNA is very similar to that of DNA, with two differences：(a) RNA contains the sugar ribose(核糖), while DNA contains the slightly different sugar deoxyribose(脱氧核糖)(a type of ribosethat lacks one oxygen atom), and (b) RNA has the nucleobase(碱基) uracil(尿嘧啶) while DNA contains thymine(胸腺嘧啶). Uracil and thymine have similar base-pairing(碱基配对) properties(特性).

RNA 与 DNA 链空间结构不同。

Unlike DNA, most RNA molecules(分子) are single-stranded(单链的). Single-stranded RNA molecules adopt very complex three-dimensional structures, since they are not restricted to(局限于) the repetitive(重复的) double-helical(双螺旋的) form of double-stranded(双链的) DNA. RNA is made within living cells by RNA polymerases(聚合酶), enzymes that act to copy a DNA or RNA template(模板) into a new RNA strand(链) through processes known as transcription(转录) or RNA replication(复制), respectively.

4.2.1 Comparison with DNA 与 DNA 的对比

RNA and DNA are both nucleic acids, but differ in three main ways. First, unlike DNA, which is, in general, double-stranded, RNA is a single-stranded molecule in many of its biological roles and has a much shorter chain of nucleotides. Second, while DNA contains deoxyribose, RNA contains ribose (in deoxyribose there is no hydroxyl group attached to the pentose ring(戊糖环) in the 2′ position). These hydroxyl groups(羟基) make RNA less stable(稳定) than DNA because it is more prone(倾向于) to hydrolysis(水解). Third, the complementary base(互补碱基) to adenine(腺嘌呤) is not thymine, as it is in DNA, but rather uracil(尿嘧啶), which is an unmethylated(未甲基化的) form of thymine.

RNA 与 DNA 的三大区别：RNA 是单链，而 DNA 为双链；RNA 糖组分为核糖，而 DNA 糖组分为脱氧核糖；RNA 中与腺嘌呤配对碱基为尿嘧啶，而 DNA 为胸腺嘧啶。

Like DNA, most biologically active RNAs, including mRNA, tRNA, rRNA, snRNAs(小核 RNA), and other noncoding(非编码的) RNAs, contain self-complementary sequences (互补序列) that allow parts of the RNA to fold(折叠) and pair with itself to form double helices(双螺旋). Analysis of these RNAs has revealed(揭示) that they are highly structured. Unlike DNA, their structures do not consist of long double helices but rather collections(聚集) of short helices(螺旋) packed together into structures akin to(近似) proteins(蛋白质). In this fashion(形状), RNAs can achieve chemical catalysis, like enzymes. For instance(例如), determination of the structure of the ribosome(核糖体)-an enzyme that catalyzes peptide bond formation-revealed that its active site(活性部位) is composed entirely of RNA.

RNA 中双螺旋结构的形成方式及特点。

RNA 的特定形状使得其能具有酶的催化能力。

4.2.2 Structure 结构

Each nucleotidein RNA contains a ribose sugar, with carbons numbered 1′ through 5′. A base is attached to the 1′ position, in general, adenine(腺嘌呤, A), cytosine(胞嘧啶, C), guanine(鸟嘌呤, G), or uracil(尿嘧啶, U). Adenine and guanine are purines(嘌呤), cytosine, and uracil are pyrimidines

RNA 的详细组成及其化学性质。

碱基配对的化学本质：碱基间的分子间作用力，如碱基间形成氢键。

(嘧啶). A phosphate group is attached to the 3′ position of one ribose(核糖) and the 5′ position of the next. The phosphate groups(磷酸基团) have a negative charge(负电荷) each at physiological pH(生理 pH), making RNA a charged(带电的) molecule (polyanion(聚阴离子)). The bases may form hydrogen bonds(氢键) between cytosine and guanine, between adenine and uracil and between guanine and uracil. However, other interactions are possible, such as a group of adenine bases binding to each other in a bulge(凸起), or the GNRA tetra-loop (四环) that has a guanine-adenine base-pair.

RNA 区别于 DNA 的一个重要结构特征在于核糖 2 号位的羟基。

2′羟基结构对 RNA 性质的影响。

An important structural feature(特征) of RNA that distinguishes it from DNA is the presence of a hydroxyl group at the 2′ position of the ribose sugar (shown below). The presence of this functional group causes the helix(螺旋) to adopt the A-form geometry(几何结构) rather than the B-form most commonly observed in DNA. This results in a very deep and narrow major groove(凹槽) and a shallow(浅的) and wide minor groove. A second consequence(结果) of the presence of the 2′- hydroxyl group is that in conformationally flexible regions(柔性构象区域) of an RNA molecule (that is, not involved in formation of a double helix), it can chemically attack the adjacent(邻近的) phosphodiester bond(磷酸二酯键) to cleave(切开) the backbone(骨架).

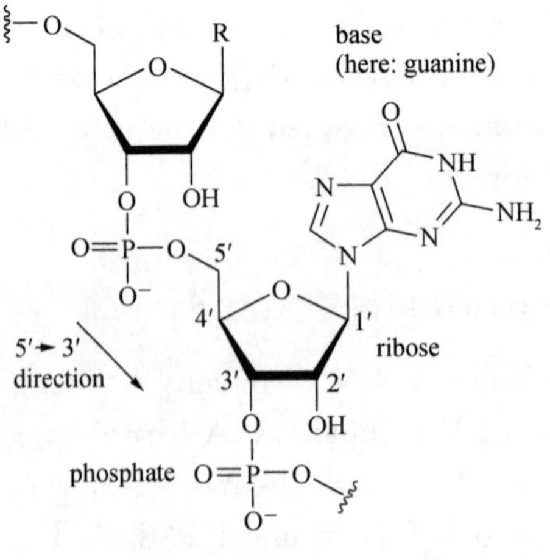

Chemical structure of RNA.

RNA is transcribed(转录) with only four bases (adenine, cytosine, guanine and uracil), but these bases and attached sugars can be modified(修饰) in numerous ways as the RNAs mature(成熟). Pseudouridine(假尿嘧啶核苷, Ψ), in which the linkage between uracil and ribose is changed from a C—N bond to a C—C bond, and ribothymidine(胸腺嘧啶核糖核苷, T) are found in various places (the most notable ones being in the TΨC loop of tRNA). Another notable modified base(修饰碱基) is hypoxanthine(次黄嘌呤), a deaminated(脱氨基) adenine base whose nucleoside(核苷) is called inosine(肌苷, I). Inosine plays a key role in the wobble hypothesis(变位假说) of the genetic code(遗传密码).

There are nearly 100 other naturally occurring modified nucleosides(核苷), of which pseudouridine and nucleosides with $2'$-O-methylribose(甲基核糖) are the most common. The specific roles of many of these modifications in RNA are not fully understood. However, it is notable that, in ribosomal RNA, many of the post-transcriptional modifications(转录后修饰) occur in highly functional regions(功能区域), such as the peptidyl transferase(肽基转移酶) center and the subunit(亚单元) interface(界面), implying(意味) that they are important for normal function.

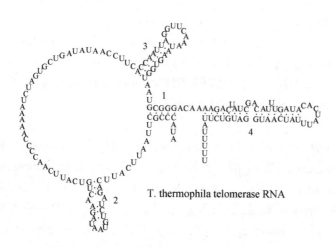

Secondary structure of a telomerase(端粒酶) RNA.

RNA 的二级结构类型。RNA 二级结构框架是由分子内氢键形成的。

The functional form of single stranded RNA molecules, just like proteins, frequently（频繁地）requires a specific tertiary structure（三级结构）. The scaffold（支架）for this structure is provided by secondary structural elements that are hydrogen bonds（氢键）within the molecule. This leads to several recognizable "domains(结构域)" of secondary structure like hairpin loops(发夹环), bulges(凸起), and internal loops（内环）. Since RNA is charged（带电的）, metalions such as Mg^{2+} are needed to stabilize(稳定) many secondary and tertiary structures.

4.2.3 Synthesis 合成

以 DNA 为模板的聚合酶催化的 RNA 合成过程称为转录。

Synthesis of RNA is usually catalyzedby an enzyme-RNA polymerase（聚合酶）-using DNA as a template, a process known as transcription(转录). Initiation(起始) of transcription begins with the binding of the enzyme to a promoter sequence（启动子序列）in the DNA（usually found "upstream(上游)" of a gene）. The DNA double helix is unwound(解旋) by the helicase(解旋酶) activity of the enzyme. The enzyme then

转录步骤。

progresses along the template strand（模板链）in the $3'$ to $5'$ direction, synthesizing a complementary（互补的）RNA molecule with elongation（延伸）occurring in the $5'$ to $3'$ direction. The DNA sequence also dictates（规定）where termination(终止) of RNA synthesis will occur. RNAs are often

RNA 经常在转录后被酶所修饰。

modified by enzymes after transcription. For example, a poly (A) tail（尾）and a $5'$ cap(头) are added to eukaryotic(真核生物的) pre-mRNA（前体信使 RNA）and introns（内含子）are removed by the spliceosome(剪接体).

有些 RNA 聚合酶是利用 RNA 为模板分子进行 RNA 合成的，如 RNA 病毒。

There are also a number of RNA-dependent RNA polymerases（聚合酶）that use RNA as their template for synthesis of a new strand of RNA. For instance, a number of RNA viruses(病毒)（such as poliovirus 脊髓灰质炎病毒）use this type of enzyme to replicate(复制) their genetic material(遗传物质). Also, RNA-dependent RNA polymerase is part of the RNA interference pathway(干扰途径) in many organisms.

4.2.4 Types of RNA 种类

(1) Overview 概述

Messenger RNA (mRNA) is the RNA that carries information from DNA to the ribosome(核糖体), the sites of protein synthesis (translation 翻译) in the cell. The coding sequence of the mRNA determines the amino acid sequence in the protein that is produced. However, many RNAs do not code for protein(about 97% of the transcriptional output is non-protein-coding in eukaryotes).

> 信使RNA简介。

These so-called non-coding(非编码的) RNAs ("ncRNA") can be encoded by their own genes (RNA genes), but can also derive from mRNA introns(内含子). The most prominent examples(突出例子) of non-coding RNAs are transfer RNA (tRNA) and ribosomal RNA (rRNA), both of which are involved in the process of translation(翻译). There are also non-coding RNAs involved in gene regulation(基因调控), RNA processing and other roles. Certain RNAs are able to catalyse chemical reactions such as cutting and ligating(结扎) other RNA molecules, and the catalysis of peptide bond formation in the ribosome; these are known as ribozymes(核酶).

> 转运RNA和核糖体RNA都是非编码RNA。
>
> 某些特定的RNA能催化化学反应,如切断或连接其他RNA分子,催化核糖体中肽键的形成,被称为核酶。

(2) In translation 翻译

Messenger RNA (mRNA) carries information about a protein sequence to the ribosomes, the protein synthesis factories in the cell. It is coded so that every three nucleotides(a codon(密码子)) correspond to one amino acid. In eukaryotic cells(真核生物细胞), once precursor(前体) mRNA(pre-mRNA) has been transcribed from DNA, it is processed to mature(成熟的) mRNA. This removes its introns-non-coding sections(非编码区 内含子) of the pre-mRNA. The mRNA is then exported(输出) from thenucleus(细胞核) to the cytoplasm(细胞质), where it is bound to ribosomes(核糖体) and translated(翻译) into its corresponding(对应的) protein form with the help of tRNA. In

> 信使RNA在翻译过程的作用。
>
> 真核生物细胞内信使RNA翻译过程。

原核生物细胞内信使RNA翻译过程。	prokaryotic cells(原核生物细胞), which do not have nucleus(细胞核) and cytoplasm(细胞质) compartments(分隔), mRNA can bind to ribosomes while it is being transcribed from DNA. After a certain amount of time the message degrades(降解) into its component nucleotides with the assistance(帮助) of ribonucleases(核糖核酸酶).
转运RNA的特点及在翻译中的作用。 转运RNA上有氨基酸结合位点及用于识别密码子的反密码子区,通过氢键结合信使RNA链上特定的序列。 核糖体RNA是核糖体中的催化组分。 核糖体结合mRNA,进行蛋白质的合成。	Transfer RNA(tRNA) is a small RNA chain of about 80 nucleotides that transfers a specific amino acid to a growing polypeptide chain(多肽链) at the ribosomal site of protein synthesis during translation. It has sites for amino acid attachment and an anticodon region(反密码子区域) for codon (密码子) recognition(识别) that binds to a specific sequence on the messenger RNA chain through hydrogen bonding. Ribosomal RNA (rRNA) is the catalytic component of the ribosomes. Eukaryotic ribosomes contain four different rRNA molecules: 18S, 5.8S, 28S and 5S rRNA. Three of the rRNA molecules are synthesized in the nucleolus, and one is synthesized elsewhere. In the cytoplasm, ribosomal RNA and protein combine to form a nucleoprotein(核蛋白) called a ribosome. The ribosome binds mRNA and carries out protein synthesis. Several ribosomes may be attached to a single mRNA at any time. rRNA is extremely abundant and makes up 80% of the 10 mg/mL RNA found in a typical eukaryotic cytoplasm.
RNA基因组的组成及其功能。	**(3) RNA genomes 基因组** Like DNA, RNA can carry genetic information. RNA viruses have genomes composed of RNA, and a variety of proteins encoded by that genome. The viral genome is replicated by some of those proteins, while other proteins protect the genome as the virus particle moves to a new host cell (宿主细胞). Viroids(类病毒) are another group of pathogens (病原体), but they consist only of RNA, do not encode any protein and are replicated by a host plant cell's polymerase.

(4) In reverse transcription 逆转录

Reverse transcribing viruses replicate their genomes by reverse transcribing DNA copies from their RNA; these DNA copies are then transcribed to new RNA. Retrotransposons(逆转录转座子) also spread by copying DNA and RNA from one another, and telomerase(端粒酶) contains an RNA that is used as template for building the ends of eukaryotic chromosomes(真核生物染色体).

逆转录病毒通过从它们的 RNA 逆转录 DNA 拷贝复制其基因组,然后这些 DNA 拷贝转录新的 RNA。

(5) Double-stranded RNA 双链的 RNA

Double-stranded RNA (dsRNA) is RNA with two complementary strands, similar to the DNA found in all cells. dsRNA forms the genetic material of some viruses (double-stranded RNA viruses). Double-stranded RNA such as viral RNA or siRNA can trigger(引发) RNA interference(干扰) in eukaryotes, as well as interferon(干扰素) response in vertebrates(脊椎动物).

双链 RNA 的概述。

4.3 DNA 脱氧核糖核酸

Deoxyribonucleic acid (DNA) is a nucleic acid that contains the genetic instructions(基因指令) used in the development and functioning of all known living organisms (with the exception of RNA viruses). The main role of DNA molecules is the long-term storage(长时间储存) of information. DNA is often compared to a set of blueprints, like a recipe or a code, since it contains the instructions needed to construct other components of cells, such as proteins and RNA molecules. The DNA segments(片段) that carry this genetic information are called genes, but other DNA sequences have structural purposes, or are involved in regulating the use of this genetic information. Along with RNA and proteins, DNA is one of the three major macromolecules that are essential for all known forms of life.

DNA 是所有已知生命形式必需的三种主要的大分子物质(其余是 RNA 和蛋白质)之一。

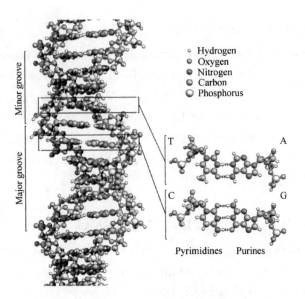

The structure of the DNA double helix.

DNA consists of two long polymers of simple units called nucleotides, with backbones made of sugars and phosphate groups joined by ester bonds(酯键). These two strands run in opposite directions(相反方向) to each other and are therefore anti-parallel(反并行). Attached to each sugar is one of four types of molecules called nucleobases (informally, bases). It is the sequence of these four nucleobases along the backbone that encodes information. This information is read using the genetic code, which specifies the sequence of the amino acids within proteins. The code is read by copying stretches of DNA into the related nucleic acid RNA in a process called transcription(转录).

Within cells DNA is organized into long structures called chromosomes(染色体). Before cell division(细胞分裂) these chromosomes are duplicated(复制) in the process of DNA replication, providing each cell its own complete set of chromosomes. Eukaryotic organisms (animals, plants, fungi, and protists(原生生物)) store most of their DNA inside the cell nucleus and some of their DNA in organelles(细胞器), such as mitochondria(线粒体) or chloroplasts(叶绿体). In contrast, prokaryotes(原核生物) (bacteria and archaea) store their DNA only in the cytoplasm(细胞浆). Within the chromosomes,

chromatin(染色质) **proteins such as histones**(组蛋白) compact and organize DNA. These compact structures(紧凑结构) guide the interactions between DNA and other proteins, helping control which parts of the DNA are transcribed.

4.3.1 Properties 特点

DNA is a long polymer made from repeating units called nucleotides. As first discovered by James D. Watson and Francis Crick, the structure of DNA of all species comprises two helical chains each coiled round the same axis(轴线), and each with a pitch of 34 Ångströms (3.4 nanometres) and a radius of 10 Ångströms (1.0 nanometres). According to another study, when measured in a particular solution, the DNA chain measured 22 to 26 Ångströms wide (2.2 to 2.6 nanometres), and one nucleotide unit measured 3.3 Å (0.33 nm) long. Although each individual repeating unit is very small, DNA polymers can be very large molecules containing millions of nucleotides. For instance, the DNA in the largest human chromosome, chromosome number 1, is approximately 220 million base-pairs long.

DNA结构首先由沃森和克里克发现。
DNA双螺旋结构的详细数据。

In living organisms DNA does not usually exist as a single molecule, but instead as a pair of molecules that are held tightly together. These two long strands entwine(纠缠) like vines(藤蔓), in the shape of a double helix. The nucleotide repeats contain both the segment of the backbone of the molecule, which holds the chain together, and a nucleobase, which interacts with the other DNA strand in the helix. A nucleobase linked to a sugar is called a nucleoside(核苷) and a base linked to a sugar and one or more phosphate groups is called a nucleotide. Polymers comprising multiple linked nucleotides (as in DNA) is called a polynucleotide(多核苷酸).

核苷酸重复单元包括碱基,戊糖及磷酸基团。

The backbone of the DNA strand is made from alternating phosphate and sugar residues. The sugar in DNA is 2-deoxyribose(2-脱氧核糖), which is a pentose (five-carbon) sugar. The sugars are joined together by phosphate groups that

组成DNA的糖是2′-脱氧核糖。

form phosphodiester bonds(磷酸二酯键) between the third and fifth carbon atoms of adjacent sugar rings(相邻糖环). These asymmetric(不对称的) bonds mean a strand of DNA has a direction. In a double helix the direction of the nucleotides in one strand is opposite to their direction in the other strand: the strands are anti-parallel. The asymmetric ends of DNA strands are called the 5′(five prime) and 3′(three prime) ends, with the 5′ end having a terminal phosphate group and the 3′ end a terminal hydroxyl group. One major difference between DNA and RNA is the sugar, with the 2-deoxyribose in DNA being replaced by the alternative pentose sugar ribose in RNA.

DNA 双螺旋由两条反并行的链组成。

Chemical structure of DNA. Hydrogen bonds shown as dotted lines.

DNA 双螺旋结构的稳定主要依靠两种作用力：核苷酸间的氢键以及芳香碱基间的碱基堆叠作用。在胞内水溶液环境中，碱基的共轭 π 键垂直于 DNA 分子轴心排列，使得其与分子溶剂化外层作用最小，而使吉布斯自由能最小。

The DNA double helix is stabilized primarily by two forces: hydrogen bonds between nucleotides and base-stacking interactions(碱基堆叠作用) among the aromatic nucleobases. In the aqueous environment of the cell, the conjugated(共轭) π bonds of nucleotide bases align perpendicular(垂直对齐) to the axis of the DNA molecule(DNA 分子轴), minimizing their interaction with the solvation shell and therefore, the Gibbs free energy(吉布斯自由能). The four bases found in DNA are adenine(腺嘌呤, abbreviated A), cytosine(胞嘧啶, C), guanine (鸟嘌呤, G) and thymine(胸腺嘧啶, T). These four bases are attached to the sugar/phosphate to form the complete nucleotide, as shown for adenosine monophosphate(单磷酸腺苷).

The nucleobases are classified into two types: the purines, A and G, being fused five-and six-membered heterocyclic compounds(杂环化合物), and the pyrimidines, the six-membered rings C and T. A fifth pyrimidine nucleobase, uracil (U), usually takes the place of thymine in RNA and differs from thymine by lacking a methyl group（甲基）on its ring. Uracil(尿嘧啶) is not usually found in DNA, occurring only as a breakdown product of cytosine(胞嘧啶). In addition to RNA and DNA a large number of artificial nucleic acid analogues（人工类核酸）have also been created to study the proprieties of nucleic acids, or for use in biotechnology.

> DNA 的四种碱基：A, G, C, T。

(1) Grooves 沟

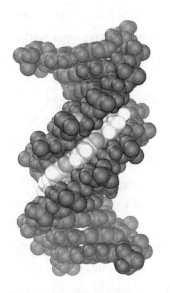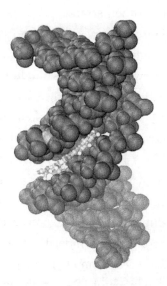

Major and minor grooves（大沟和小沟）of DNA. Minor groove is a binding site for the dye(染料) Hoechst 33258.

双螺旋链形成 DNA 骨架。
DNA 沟的位置及作用。

DNA 沟的尺寸。
狭窄的小沟意味着碱基的边缘更容易进入大沟。

Twin helical strands form the DNA backbone. Another double helix may be found by tracing the spaces, or grooves, between the strands. These voids(空隙) are adjacent to the base pairs and may provide a binding site(结合位点). As the strands are not directly opposite each other, the grooves are unequally sized. One groove, the major groove, is 22 Å wide and the other, the minor groove, is 12 Å wide. The narrowness of the minor groove means that the edges of the bases are more accessible in the major groove. As a result, proteins like transcription factors that can bind to specific sequences in double-stranded DNA usually make contacts to the sides of the bases exposed in the major groove. This situation varies in unusual conformations of DNA within the cell, but the major and minor grooves are always named to reflect the differences in size that would be seen if the DNA is twisted(扭曲的) back into the ordinary B form.

(2) Base pairing 碱基配对

DNA 中碱基配对规则：A 与 T，C 与 G。
碱基对的定义。

DNA 双螺旋中的两条链可以像拉开拉链一样，通过机械力或高温打开。

互补碱基间的可逆和特定的作用对生物体中 DNA 的所有功能至关重要。

In a DNA double helix, each type of nucleobase on one strand normally interacts with just one type of nucleobase on the other strand. This is called complementary base pairing. Here, purines form hydrogen bonds to pyrimidines, with A bonding only to T, and C bonding only to G. This arrangement of two nucleotides binding together across the double helix is called a base pair. As hydrogen bonds are not covalent(共价键), they can be broken and rejoined relatively easily. The two strands of DNA in a double helix can therefore be pulled apart like a zipper(拉链), either by a mechanical force(机械力) or high temperature. As a result of this complementarity, all the information in the double-stranded sequence of a DNA helix is duplicated(复制) on each strand, which is vital in DNA replication. Indeed, this reversible and specific interaction between complementary base pairs is critical for all the functions of DNA in living organisms.

A GC base pair with three hydrogen bonds.

An AT base pair with two hydrogen bonds. Non-covalent hydrogen bonds between the pairs are shown as dashed lines(虚线).

The two types of base pairs form different numbers of hydrogen bonds, AT forming two hydrogen bonds, and GC forming three hydrogen bonds (see figures above). DNA with high GC-content is more stable than DNA with low GC-content. Although it is often stated that this is due to the added stability of an additional hydrogen bond, this is incorrect. DNA with high GC-content is more stable due to intra-strand base stacking interactions(链内碱基堆叠作用).

高 GC 含量的 DNA 比低 GC 含量的 DNA 更稳定。虽然常说这是由于额外的一个氢键增加了稳定性,然而这种说法是不正确的。高 GC 含量的 DNA 的稳定性增加是由于链内碱基堆叠作用。

As noted above, most DNA molecules are actually two polymer strands, bound together in a helical fashion by noncovalent bonds; this double stranded structure (dsDNA) is maintained largely by the intra-strand base stacking interactions, which are strongest for G,C stacks. The two strands can come apart—a process known as melting—to form two single-stranded DNA(ssDNA) molecules. Melting occurs when conditions favor ssDNA; such conditions are high temperature, low salt and high pH (low pH also melts DNA,

DNA 双链分离称作融化,常用条件包括:高温,低盐,高 pH。

but since DNA is unstable due to acid depurination(脱嘌呤作用), low pH is rarely used).

The stability(稳定性) of the dsDNA form depends not only on the GC-content (% G, C base pairs) but also on sequence (since stacking is sequence specific) and also length (longer molecules are more stable). The stability can be measured in various ways; a common way is the "melting temperature(熔点)", which is the temperature at which 50% of the ds molecules are converted to ss molecules; melting temperature is dependent on ionic strength(离子强度) and the concentration of DNA. As a result, it is both the percentage of GC base pairs and the overall length of a DNA double helix that determine the strength of the association between the two strands of DNA. Long DNA helices with a high GC-content have stronger-interacting strands, while short helices with high AT content have weaker-interacting strands.

In the laboratory, the strength of this interaction can be measured by finding the temperature required to break the hydrogen bonds, their melting temperature (also called Tm value). When all the base pairs in a DNA double helix melt, the strands separate and exist in solution as two entirely independent molecules. These single-stranded DNA molecules (ssDNA) have no single common shape, but some conformations(构象) are more stable than others.

(3) Sense and antisense 正义和反义

A DNA sequence is called "sense"(正义) if its sequence is the same as that of a messenger RNA copy that is translated into protein. The sequence on the opposite strand is called the "antisense" sequence. Both sense and antisense sequences can exist on different parts of the same strand of DNA (i.e. both strands contain both sense and antisense sequences). In both prokaryotes(原核生物) and eukaryotes(真核生物), antisense RNA sequences are produced, but the functions of these RNAs are not entirely clear. One proposal is that antisense RNAs are involved in regulating gene expression(表达调控) through RNA-RNA base pairing.

A few DNA sequences in prokaryotes and eukaryotes, and more in plasmids(质粒) and viruses(病毒), blur(模糊) the distinction between sense and antisense strands by having overlapping genes(重叠基因). In these cases, some DNA sequences do double duty, encoding one protein when read along one strand, and a second protein when read in the opposite direction along the other strand. In bacteria, this overlap may be involved in the regulation of gene transcription, while in viruses, overlapping genes increase the amount of information that can be encoded within the small viral genome(基因组).

(4) Supercoiling 超螺旋

DNA can be twisted(扭曲) like a rope(绳) in a process called DNA supercoiling. With DNA in its "relaxed" state, a strand usually circles the axis of the double helix once every 10.4 base pairs, but if the DNA is twisted the strands become more tightly or more loosely wound. If the DNA is twisted in the direction of the helix, this is positive supercoiling, and the bases are held more tightly together. If they are twisted in the opposite direction, this is negative supercoiling(负超螺旋), and the bases come apart more easily. In nature, most DNA has slight negative supercoiling that is introduced by enzymes(酶) called topoisomerases(拓扑异构酶). These enzymes are also needed to relieve the twisting stresses introduced into DNA strands during processes such as transcription and DNA replication.

(5) Alternate DNA structures 交替的DNA结构

DNA exists in many possible conformations that include A-DNA, B-DNA, and Z-DNA forms, although, only B-DNA and Z-DNA have been directly observed in functional organisms. The conformation that DNA adopts depends on the hydration level(水合水平), DNA sequence, the amount and direction of supercoiling, chemical modifications(化学修饰) of the bases, the type and concentration of metal ions, as well as the presence of polyamines(多胺) in solution.

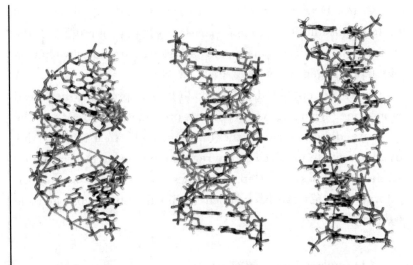

From left to right, the structures of A, B and Z DNA.

A-DNA 和 B-DNA 的比较。

Compared to B-DNA, the A-DNA form is a wider right-handed spiral(右手螺旋), with a shallow, wide minor groove and a narrower, deeper major groove. The A form occurs under non-physiological conditions(非生理条件) in partially dehydrated(部分脱水) samples of DNA, while in the cell it may be produced in hybrid pairings of DNA and RNA strands, as well as in enzyme-DNA complexes(复合物). Segments of DNA where the bases have been chemically modified by methylation(甲基化) may undergo a larger change in conformation and adopt the Z form. Here, the strands turn about the helical axis in a left-handed spiral(左手螺旋), the opposite of the more common B form. These unusual structures can be recognized by specific Z-DNA binding proteins and may be involved in the regulation of transcription.

(6) Base modifications 碱基修饰

基因表达的运行机制。

The expression of genes(基因表达) is influenced by how the DNA is packaged in chromosomes(染色体), in a structure called chromatin(染色质). Base modifications can be involved in packaging, with regions that have low or no gene expression usually containing high levels of methylation of cytosine bases

碱基甲基化。

(胞嘧啶碱基的甲基化). For example, cytosine methylation(胞

嘧啶甲基化），produces 5-methylcytosine, which is important for X-chromosome inactivation(失活).

Despite the importance of 5-methylcytosine, it can deaminate(脱氨基) to leave a thymine base, so methylated cytosines are particularly prone to mutations(突变).

5-甲基胞嘧啶的生物学意义。

4.3.2　Biological functions 生物功能

DNA usually occurs as linear chromosomes in eukaryotes, and circular chromosomes in prokaryotes. The set of chromosomes in a cell makes up its genome; the human genome has approximately 3 billion base pairs of DNA arranged into 46 chromosomes. The information carried by DNA is held in the sequence of pieces of DNA called genes. Transmission of genetic information in genes is achieved via complementary base pairing. For example, in transcription, when a cell uses the information in a gene, the DNA sequence is copied into a complementary RNA sequence through the attraction between the DNA and the correct RNA nucleotides. Usually, this RNA copy is then used to make a matching protein sequence in a process called translation, which depends on the same interaction between RNA nucleotides. In alternative fashion, a cell may simply copy its genetic information in a process called DNA replication.

DNA通常出现在真核生物的线性染色体中以及原核生物的环状染色体中。

基因的遗传通过碱基互补配对。

【Key words】

nucleic acid 核酸	phosphate 磷酸
ribonucleic acid (RNA) 核糖核酸	purine 嘌呤
deoxyribonucleic acid (DNA) 脱氧核糖核酸	pyrimidine 嘧啶
nucleoside 核苷	3′,5′-phosphodiester bond 3′,5′-磷酸二酯键
nucleotide 核苷酸	base-stacking interaction 碱基堆砌作用
base 碱基	base pairing 碱基配对
pentose 戊糖	supercoiling 超螺旋

Questions

1. If the content of base A is 15% in a double strand DNA molecule, that of base C is (　　)
   A. 15%　　　　B. 35%　　　　C. 75%　　　　D. 85%
2. The linked bond between nucleotides is _____ in nucleic acids.
3. The temperature at which half of the DNA molecules have denatured is called _____ for that DNA.
4. The wavelength at characteristic absorption peak of nucleic acids is _____ nm.
5. What is the base pairing complementary principle?
6. Compare the structural differences between RNA and DNA.

References

[1] McKee Trudy, R McKee James. Biochemistry: An Introduction (Second Edition)[M]. New York: McGraw-Hill Companies, 1999.

[2] Nelson D L, Cox M M. Lehninger Principles of Biochemistry (Third Edition)[M]. Derbyshire: Worth Publishers, 2000.

[3] Ding W, Jia H T. Biochemistry (First Edition)[M]. Beijing: Higher Education Press, 2012.

[4] 郑集, 陈钧辉. 普通生物化学(第四版)[M]. 北京:高等教育出版社, 2007.

[5] 王艳萍. 生物化学[M]. 北京:中国轻工业出版社, 2013.

Chapter 5　Enzyme 酶

Enzymes are macromolecular biological catalysts(催化剂), which are known to catalyze more than 5,000 biochemical reaction types. Enzymes accelerate(加速), or catalyze, chemical reactions. The molecules at the beginning of the process upon which enzymes may act are called substrates(底物) and the enzyme converts these into different molecules, called products(产物). Almost all metabolic processes(代谢过程) in the cell need enzymes in order to occur at rates fast enough to sustain life.

> 酶与底物作用,加速生化反应。

Most enzymes are proteins, although a few are catalytic RNA molecules. Like all catalysts, enzymes increase the rate of a reaction by lowering its activation energy(活化能). Chemically, enzymes are like any catalyst and are not consumed(消耗) in chemical reactions, nor do they alter the equilibrium(平衡) of a reaction. Enzymes differ from most other catalysts by being much more specific. Enzyme activity can be affected by other molecules: inhibitors(抑制剂) are molecules that decrease enzyme activity, and activators(激活剂) are molecules that increase activity.

> 酶与一般催化剂的异同点。

Some enzymes are used commercially(商业化), for example, in the synthesis of antibiotics(抗生素). Some household products use enzymes to speed up chemical reactions: enzymes in biological washing powders break down protein, starch(淀粉) or fat stains(脂肪污渍) on clothes, and enzymes in meat tenderizer(嫩肉粉) break down proteins into smaller molecules, making the meat easier to chew(咀嚼).

> 酶的一些商业用途。

5.1 Enzyme Structure 酶结构

Enzymes are generally globular proteins(球形蛋白质), acting alone or in larger complexes(复合物). Like all proteins, enzymes are linear chains of amino acids that fold to produce a three-dimensional(三维) structure. The sequence of the amino acids specifies the structure which in turn determines the catalytic activity(催化活性) of the enzyme. Although structure determines function, a novel enzyme's activity can not yet be predicted from its structure alone. Enzyme structures unfold (denature) when heated or exposed to chemical denaturants(变性剂) and this disruption to the structure typically causes a loss of activity. Enzyme denaturation(变性) is normally linked to temperatures above a species' normal level.

Enzymes are usually much larger than their substrates. Only a small portion of their structure (around 2～4 amino acids) is directly involved in catalysis: the catalytic site(催化部位). This catalytic site is located next to one or more binding sites(结合部位) where residues(残基) orient the substrates. The catalytic site and binding site together comprise the enzyme's active site(活性中心). The remaining majority of the enzyme structure serves to maintain the precise orientation(定位) and dynamics(动力学) of the active site. Enzyme structures may also contain allosteric sites(别构部位) where the binding of a small molecule causes a conformational(构象) change that increases or decreases activity.

A small number of RNA-based biological catalysts called ribozymes(核酶) exist, which again can act alone or in complex with proteins. The most common of these is the ribosome which is a complex of protein and catalytic RNA components.

大部分酶为蛋白质,它们的一级结构决定高级结构,进而决定其催化活性。酶结构破坏,活性也会被破坏。

酶的活性中心。

酶活性中心外的结构对于维持酶的活性中心很重要,有的酶具有别构部位,可以与其他分子作用,改变酶的构象及活性。

少数酶为 RNA,称为核酶。

5.2　Enzyme catalysis 酶催化反应

The mechanism(机理) of enzyme catalysis is similar in principle(本质上) to other types of chemical catalysis. By providing an alternative reaction route and by stabilizing intermediates(稳定中间体), the enzyme reduces the energy required to reach the highest energy transition state(过渡状态) of the reaction. The reduction of activation energy(E_a)(活化能) increases the number of reactant molecules(反应分子) with enough energy to reach the activation energy and form the product.

酶促反应本质。

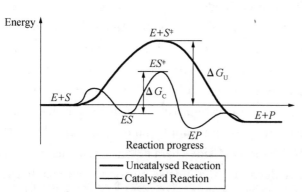

Stabilization of the transition state by an enzyme.

Enzymes must bind their substrates(底物) before they can catalyze any chemical reaction. Enzymes are usually very specific as to what substrates they bind and then the chemical reaction catalyzed. Specificity(专一性) is achieved by binding pockets with complementary(互补的) shape, charge(电荷) and hydrophilic/hydrophobic(亲水的/疏水的) characteristics(特点) to the substrates. Enzymes can therefore distinguish(区别) between very similar substrate molecules to be chemoselective(化学选择性), regioselective(区域选择性) and stereospecific(立体特异性). Some of the enzymes showing the highest specificity and accuracy(准确性). Conversely, some enzymes have broad specificity and act on a range of different physiologically(生理学上) relevant substrates.

酶与底物结合的专一性。

绝对专一性与相对专一性。

To explain the observed specificity of enzymes, two binding models of enzyme and substrate have been proposed.

(1) "Lock and key" model 锁钥学说

In 1894, Emil Fischer proposed that both the enzyme and the substrate possess specific complementary geometric shapes (互补的几何形状) that fit exactly into one another. This is often referred to as the "lock and key" model. This early model explains enzyme specificity, but fails to explain the stabilization of the transition state(稳定过渡态) that enzymes achieve.

(2) Induced fit model 诱导契合学说

The favored model for the enzyme-substrate interaction is the induced fit model. In 1958, Daniel Koshland suggested a modification to the lock and key model: since enzymes are rather flexible(灵活的) structures, the active site(活性中心) is continuously reshaped(重塑) by interactions with the substrate (底物) as the substrate interacts with the enzyme. As a result, the substrate does not simply bind to a rigid(刚性的) active site; the amino acid side-chains(侧链) that make up the active site are molded into the precise positions that enable the enzyme to perform its catalytic function.

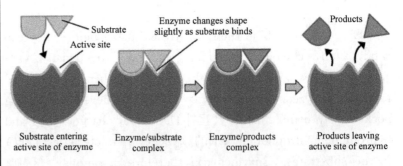

Diagrams to show the induced fit hypothesis of enzyme action.

5.3 Enzyme kinetics 酶动力学

The reaction catalyzed(催化) by an enzyme uses exactly the same reactants and produces exactly the same products as the unanalyzed reaction. Like other catalysts(催化剂), enzymes do not alter(改变) the position of equilibrium(平衡) between substrates and products. However, unlike unanalyzed chemical reactions, enzyme-catalyzed reactions display saturation kinetics.

For a given enzyme concentration(浓度) and for relatively low substrate concentrations, the reaction rate increases linearly with substrate concentration; the enzyme molecules are largely free to catalyze the reaction, and increasing substrate concentration means an increasing rate at which the enzyme and substrate molecules encounter one another. However, at relative high substrate concentrations, the reaction rate asymptotically(渐近地) approaches(达到) the theoretical(理论的) maximum; the enzyme active sites are almost all occupied and the reaction rate is determined by the intrinsic(固有) turnover(周转) rate of the enzyme.

与没有催化的反应相比，酶催化条件下对反应物的使用量和生成产物的量没有影响。与其他催化剂一样，酶不改变底物与产物之间的平衡位置。

底物浓度较低，酶浓度一定的情况下，反应速率随底物浓度地增加呈线性增加；酶在很大程度上自由地催化反应，增加底物浓度意味着增加酶与底物分子相遇的速率。但是，在底物浓度较高的情况下，反应速度趋于理论上的最大化，因为酶的活性中心几乎都被占据，而反应速率取决于酶的固有周转速率。

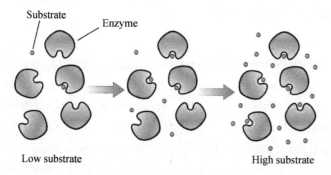

动画：酶动力学

Reaction rates increase as substrate concentration increase, but become saturated at very high concentrations of substrate.

Enzyme kinetics is a discipline to study the rate of enzymatic reaction and how it changes in the response to the factors influencing enzyme action, such as substrate concentration, enzyme concentration, pH, temperature, inhibitor and activator. The rate data used in kinetic analyses are commonly obtained from enzyme assays.

5.3.1 Enzyme assays 酶测定法

Enzyme assays are laboratory procedures that measure the rate of enzyme reactions. Because enzymes are not consumed by the reactions they catalyze, enzyme assays usually follow changes in the concentration of either substrates or products to measure the rate of reaction. There are many methods of measurement.

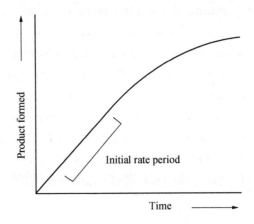

Progress curve for an enzyme reaction. The slope in the initial rate period is the initial rate of reaction v.

Spectrophotometric assays（分光光度分析）observe change in the absorbance of light between products and reactants; radiometric assays（放射分析）involve the incorporation（结合）or release of radioactivity（放射性）to measure the amount of product made over time. Spectrophotometric assays are most convenient（方便）since they allow the rate of the reaction to be measured continuously（连续地）. Although radiometric assays require the removal and counting of samples (i.e., they are discontinuous assays) they are usually extremely sensitive（灵敏）and can measure very low levels of enzyme activity. An analogous approach（类似的方法）is to use mass spectrometry

（质谱分析）to monitor the incorporation（掺入）or release of stable isotopes（同位素）as substrate is converted into product.

The most sensitive enzyme assays use lasers（激光）focused through a microscope to observe changes in single enzyme molecules as they catalyze their reactions. These measurements either use changes in the fluorescence（荧光）of cofactors（辅助因子）during an enzyme's reaction mechanism, or of fluorescent dyes（荧光染料）added onto specific sites of the protein to report movements that occur during catalysis. These studies are providing a new view of the kinetics and dynamics of single enzymes, as opposed to traditional enzyme kinetics, which observes the average behavior of populations of millions of enzyme molecules.

An example progress curve（曲线）for an enzyme assay is shown above. The enzyme produces product at an initial rate（初速度）that is approximately linear for a short period after the start of the reaction. As the reaction proceeds and substrate is consumed, the rate continuously slows (so long as substrate is not still at saturating levels（饱和的）). To measure the initial (and maximal) rate, enzyme assays are typically carried out while the reaction has progressed only a few percent towards total completion. The length of the initial rate period depends on the assay conditions and can range from milliseconds（毫秒）to hours. However, equipment for rapidly mixing liquids allows fast kinetic measurements on initial rates of less than one second.

Most enzyme kinetics studies concentrate on this initial, approximately linear part of enzyme reactions. However, it is also possible to measure the complete reaction curve and fit this data to a non-linear rate equation. This way of measuring enzyme reactions is called progress-curve analysis. This approach is useful as an alternative to rapid kinetics when the initial rate is too fast to measure accurately.

5.3.2　Michaelis–Menten kinetics 米-曼氏动力学

In 1913, Leonor Michaelis and Maud Leonora Menten proposed(提出) a quantitative theory of enzyme kinetics, which describes the effect of substrate concentration(底物浓度) on enzymatic reaction. The major contribution of Michaelis and Menten was to think of enzyme reactions in two stages. In the first, the substrate binds reversibly to the enzyme, forming the enzyme – substrate complex(酶-底物复合物). This is sometimes called the Michaelis-Menten complex in their honor. The enzyme then catalyzes the chemical step in the reaction and releases the product. This work was further developed by G. E. Briggs and J. B. S. Haldane, who derived kinetic equations that are still widely used today.

> 酶催化反应是可以饱和的,其反应速率并不随着底物的增加呈线性变化。反应初速度的测定是通过检测一定范围内的底物浓度(表示[S]),此时反应速率(v)随着[S]增加而增加。然而,随着[S]增高,酶逐渐被底物饱和,速度达到酶的最大速率V_{max}。

As enzyme-catalyzed reactions are saturable(可饱和的), their rate of catalysis does not show a linear response to increasing substrate. The initial rate of the reaction is measured over a range of substrate concentrations (denoted as [S]), the reaction rate (v) increases as [S] increases, as shown below. However, as [S] gets higher, the enzyme becomes saturated with substrate and the rate reaches V_{max}, the enzyme's maximum rate.

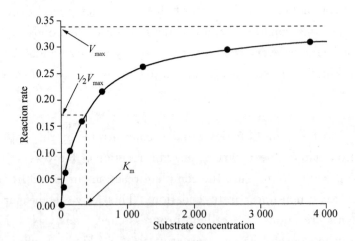

Saturation curve for an enzyme showing the relation between the concentration of substrate and rate.

$$E+S \underset{k_{-1}}{\overset{k_1}{\rightleftharpoons}} ES \xrightarrow{k_2} E+P$$

Substrate binding　　Catalytic step

Single-substrate mechanism for an enzyme reaction. k_1, k_{-1} and k_2 are the rate constants for the individual steps.

The Michaelis—Menten kinetic model of a single-substrate reaction is shown above. There is an initial bimolecular reaction between the enzyme E and substrate S to form the enzyme-substrate complex ES. Although the enzymatic mechanism for the unimolecular(单分子) reaction $ES \xrightarrow{k_{cat}} E+P$ can be quite complex, there is typically one rate-determining enzymatic step that allows this reaction to be modelled as a single catalytic step with an apparent unimolecular rate constant(单分子速率常数) k_{cat}. If the reaction path proceeds over one or several intermediates(中间体), k_{cat} will be a function of several elementary rate constants(初级速率常数), whereas in the simplest case of a single elementary reaction (e. g. no intermediates), it will be identical to the elementary unimolecular rate constant k_2. The apparent unimolecular rate constant k_{cat} is also called turnover number(周转次数) and denotes(表示) the maximum number of enzymatic reactions catalyzed per second.

The Michaelis-Menten equation describes how the initial reaction rate v_0 depends on the position of the substrate-binding equilibrium(平衡) and the rate constant k_2.

$$v_0 = \frac{V_{max}[S]}{K_M + [S]} \text{ (Michaelis-Menten equation)}$$

with the constants(常数):

$$K_M \stackrel{def}{=\!=} \frac{k_2 + k_{-1}}{k_1} \approx K_D$$

$$V_{max} \stackrel{def}{=\!=} k_{cat}[E]_{tot}$$

This Michaelis-Menten equation is the basis for most single-substrate enzyme kinetics. Two crucial assumptions(假设) underlie

单底物反应的米-曼氏动力学模型见上图。酶 E 与底物 S 最开始先形成一个双分子的酶-底物复合物 ES。虽然单分子酶促反应机制很复杂,但是通常有一个是决定速度的酶步骤,使其可以模拟成具有一个表观单分子速率常数 k_{cat} 的单步骤酶促反应。如果反应途径存在一个或多个中间体,k_{cat} 则是几个初级速率常数的功能体。

对于最简单的单分子反应(没有中间体)而言,初级速率常数 k_2 相同。表观单分子速率常数 k_{cat} 也被称为周转次数,表示每秒钟最大酶促反应次数。

米-曼氏方程描述了初始反应速率 v_0 如何取决于底物结合的平衡位置以及速率常数 k_2。

米-曼氏方程

米-曼氏方程是大多数单底物酶动力学的基础。该方程基于两个关键假设(除了该机制的一般假设即没有中间体或产物抑制,而且没有变构或协同作用之外)。

this equation (apart from the general assumption about the mechanism only involving no intermediate or product inhibition, and there is no allostericity(变构) or cooperativity(协同)).

第一个假设是所谓的准稳态假设(或假稳态假说),即底物结合的酶浓度变化比产物和底物的变化慢很多。

第二个假设是,总的酶浓度不随时间变化。

The first assumption is the so called quasi-steady-state assumption (or pseudo-steady-state hypothesis), namely that the concentration of the substrate-bound enzyme (and hence also the unbound enzyme) changes much more slowly than those of the product and substrate and thus the change over time of the complex can be set to zero $d[ES]/dt \overset{!}{=} 0$. The second assumption is that the total enzyme concentration does not change over time, thus $[E]_{tot}=[E]+[ES] \overset{!}{=} \text{const}$.

米氏常数 K_M 被定义为反应速率等于最大反应速率一半时的底物浓度。

The Michaelis constant K_M is experimentally defined as the concentration at which the rate of the enzyme reaction is half V_{max}, which can be verified(证实) by substituting(代替) $[S]=K_M$ into the Michaelis-Menten equation and can also be seen graphically. If the rate-determining enzymatic step is slow compared to substrate dissociation(解离) ($k_2 \ll k_{-1}$), the Michaelis constant K_M is roughly(大致) the dissociation constant(解离常数) K_D of the ES complex.

If $[S]$ is small compared to K_M then the term $[S]/([K_M]+[S]) \approx [S]/K_M$ and also very little ES complex is formed, thus $[E]_0 \approx [E]$. Therefore, the rate of product formation is

$$v_0 \approx \frac{k_{cat}}{K_M}[E][S] \qquad \text{if } [S] \ll K_M$$

产物生产速率取决于酶的浓度以及底物浓度,方程类似于具有相应假二级速率常数 K_2/K_M 的双分子反应,这个常数可用于衡量催化效率。最高效的酶的 K_2/K_M 范围为 $10^8 \sim 10^{10}$ $M^{-1} \cdot s^{-1}$。

Thus the product formation rate depends on the enzyme concentration as well as on the substrate concentration, the equation resembles a bimolecular reaction with a corresponding pseudo-second order rate constant K_2/K_M. This constant is a measure of catalytic efficiency(催化效率). The most efficient enzymes reach a K_2/K_M in the range of $10^8 \sim 10^{10}$ $M^{-1} \cdot s^{-1}$. These enzymes are so efficient they effectively catalyze a reaction each time they encounter a substrate molecule and have thus reached an upper theoretical limit(理论上限) for efficiency (diffusion limit)（扩散限制）; these enzymes have often been termed perfect enzymes.

5.3.3 Linear plots of the Michaelis-Menten equation 米-曼氏方程的线性作图法

Using an interactive Michaelis-Menten kinetics tutorial（教程）at the University of Virginia, the effects on the behaviour of an enzyme of varying kinetic constants can be explored.

The plot of v versus $[S]$ above is not linear（线性的）; although initially linear at low $[S]$, it bends over to saturate at high $[S]$. Before the modern era of nonlinear curve-fitting（曲线拟合）on computers, this nonlinearity could make it difficult to estimate K_M and V_{max} accurately. Therefore, several researchers developed linearisations（线性化）of the Michaelis-Menten equation, such as the Lineweaver-Burk plot, the Eadie-Hofstee diagram and the Hanes-Woolf plot. All of these linear representations can be useful for visualizing（可视化）data, but none should be used to determine kinetic parameters（参数）, as computer software is readily available that allows for more accurate determination by nonlinear regression methods（非线性回归方法）.

在具有现代非线性曲线拟合功能的计算机诞生之前，这种非线性曲线很难准确地计算出米氏常数 K_M 和最大反应速度 V_{max}。因此，一些研究人员对米-曼氏方程进行了线性化开发。

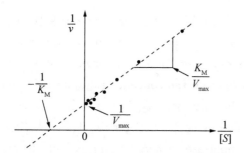

Lineweaver-Burk or double-reciprocal plot of kinetic data, showing the significance of the axis intercepts and gradient.

The Lineweaver-Burk plot or double reciprocal plot is a common way of illustrating kinetic data. This is produced by taking the reciprocal（倒数）of both sides of the Michaelis-Menten equation. As shown above, this is a linear form of the

Lineweaver-Burk 图或双倒数作图是用于分析酶动力学数据的一种常见方式。

Michaelis-Menten equation and produces a straight line with the equation(方程式) $y=mx+c$ with a y-intercept equivalent to $1/V_{max}$ and an x-intercept of the graph representing $-1/K_M$.

$$\frac{1}{v} = \frac{K_M}{V_{max}[S]} + \frac{1}{V_{max}}$$

Lineweaver-Burk 图或双倒数作图的方程式。

Naturally, no experimental values can be taken at negative $1/[S]$; the lower limiting value $1/[S]=0$ (the y-intercept) corresponds to an infinite substrate concentration, where $1/v=1/V_{max}$ as shown above; thus, the x-intercept (x 轴上的截距) is an extrapolation of the experimental data taken at positive concentrations. More generally, the Lineweaver-Burk plot skews the importance of measurements taken at low substrate concentrations and, thus, can yield inaccurate estimates of V_{max} and K_M.

5.3.4 Significance of kinetic constants 动力学常数的意义

研究动力学常数的两大意义。

The study of enzyme kinetics is important for two basic reasons. Firstly, it helps explain how enzymes work, and secondly, it helps predict(预测) how enzymes behave in living organisms. The kinetic constants defined above, K_M and V_{max}, are critical to attempts to understand how enzymes work together to control metabolism.

解释最大反应速度 V_{max} 与米-曼氏常数 K_M 的含义。

The Michaelis-Menten equation(米-曼氏方程) describes the kinetic behavior of a great many enzymes. V_{max} is an important characteristic of a given enzyme-substrate system and is dependent only on the concentration and turnover number(周转次数) of the enzyme. K_M is the substrate concentration at which the reaction rate is half-maximal(最大反应速度的一半), or K_M is the substrate concentration at which the enzyme is half-saturated(一半被饱和) with its substrate. For most enzymes, the K_M values lies between 10^{-1} and 10^{-7} mol/L. The K_M value is a characteristic constant(特征性常数) of enzymes. The K_M value for an enzyme depends on the particular substrate(特定的底物) and on the environmental conditions, such as temperature, pH, and ionic strength(离子强度), regardless of enzyme concentrations. The smaller K_M value, the larger the affinity(亲和力) of the enzyme, and the faster the enzyme reaction.

5.3.5 Enzyme concentrations 酶浓度

In addition to the effect of substrate concentration, enzymatic reactions are dependent on the concentration of enzyme. The initial velocity(初速度) of an enzymatic reaction is directly proportional to(成正比) the concentration of enzyme, provided that the substrate is present in excess.

底物浓度足够大的前提下,酶反应速度与酶浓度成正比。

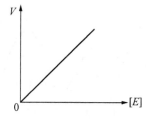

Effect of enzyme concentrations on enzymatic velocity.

5.3.6 Temperature and pH 温度与 pH

Enzymes have an optimum temperature(最适温度) at which their activity is maximal. Enzymatic reactions are accelerated by increasing the temperature until an optimum value is reached; thereafter, the reaction declines. This because of that greater activation of energy is required to an enzymatic reaction under a lower temperature; at excessively high temperatures, however, enzyme becomes irreversibly denatured and lost its catalytic activity.

温度对酶反应速度的影响。

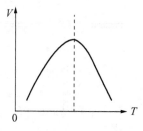

Temperature dependence of enzymatic reaction.

微课:影响酶促反应的因素

pH 对酶反应速度的影响。

Enzymes have an optimum pH (or pH range) at which their activity is maximal; at higher or lower pH, activity decreases. This is not surprising, because the pH would conceivably affect the binding of enzyme to substrate and also functioning of the active sites. Amino acid side chains in the active site(活性中心) may act as weak acids and bases with critical functions that depend on their maintaining a certain state of ionization(电离作用), and elsewhere(在别处) in the protein ionized side chains(侧链) may play an essential role in the interactions that maintain protein structure.

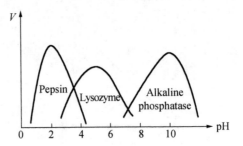

pH dependence of some enzymes.

5.3.7　Enzyme inhibition and activation 酶抑制与激活

酶抑制剂的特点。

Enzyme inhibitors(抑制剂) are molecules that reduce or abolish(废止) enzyme activity, while enzyme activators are molecules that increase the catalytic(催化) rate of enzymes. These interactions can be either reversible(可逆的)(i.e., removal of the inhibitor restores enzyme activity) or irreversible (i.e., the inhibitor permanently(永久) inactivates(使失活) the enzyme).

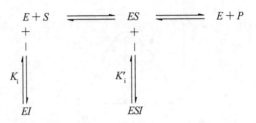

Kinetic scheme for reversible enzyme inhibitors.

Reversible inhibitors 可逆性抑制剂

Traditionally reversible enzyme inhibitors have been classified as competitive(竞争性), noncompetitive(非竞争性), or uncompetitive(反竞争性), according to their effects on K_M and V_{max}. These different effects result from the inhibitor binding to the enzyme E, to the enzyme-substrate complex ES, or to both, respectively.

> 可逆性抑制剂的分类。

The particular type of an inhibitor can be discerned(区分) by studying the enzyme kinetics as a function of the inhibitor concentration. The three types of inhibition produce Lineweaver-Burke plots that vary in distinctive(独特的) ways with inhibitor concentration.

(1) **A competitive inhibitor** competes with the substrate for the active site(活性中心) of an enzyme. While the inhibitor (I) occupies the active site it prevents binding of the substrate(底物) to the enzyme. Many competitive inhibitors are compounds(复合物) that resemble(类似于) the substrate and combine with the enzyme to form an EI complex, but without leading to catalysis(催化). Even fleeting(快速过去的) combinations of this type will reduce the efficiency of the enzyme. However, when $[S]$ far exceeds $[I]$, the probability(可能性) that an inhibitor molecule will bind to the enzyme is minimized(最小化) and the reaction exhibits a normal V_{max}. In the Lineweaver-Burke plot, compared with no inhibitor, the apparent K_M(表观 K_M) is increased, while the V_{max} is unchanged.

> 竞争性抑制的特点。

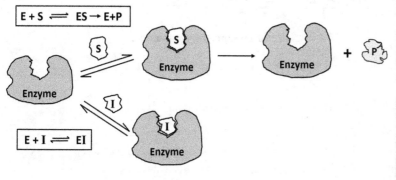

Competitive inhibition.

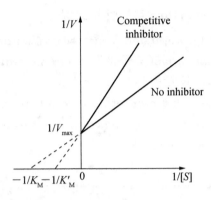

Lineweaver-Burke plot showing competitive inhibition.

非竞争性抑制的特点。

(2) **Noncompetitive inhibition** is generally characterized as an inhibition of enzymatic activity by compounds bearing no structural relationship to the substrate, and therefore the inhibition cannot be reversed(逆转) by increased concentration of substrate. Unlike competitive inhibitors, noncompetitive inhibitors cannot interact at the active site but must bind to some other portion of an enzyme or enzyme-substrate complex. In the Lineweaver-Burke plot, compared with no inhibitor, the apparent K_M is unchanged, while the V_{max} is decreased.

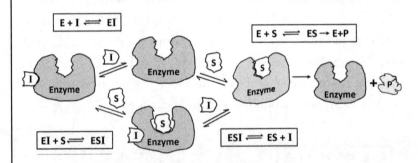

Noncompetitive inhibition.

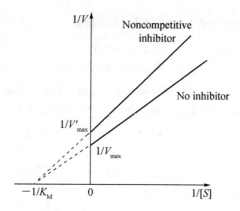

Lineweaver-Burke plot showing noncompetitive inhibition.

(3) Another form of reversible inhibition is called **uncompetitive inhibition**, in which the uncompetitive inhibitors bind only to the enzyme-substrate complex ES. In the Lineweaver-Burke plot, both apparent K_M and V_{max} are decreased.

反竞争性抑制的特点。

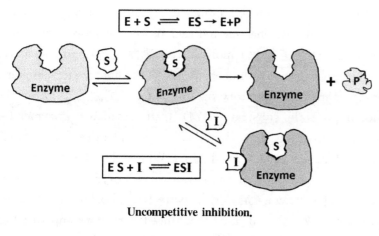

Uncompetitive inhibition.

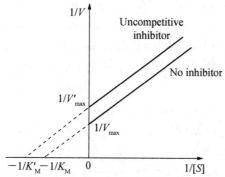

Lineweaver-Burke plot showing uncompetitive inhibition.

Irreversible inhibitors 不可逆抑制剂

Enzyme inhibitors can also irreversibly inactivate enzymes, usually by covalently modifying(修改) active site residues(活性中心残基). Farm chemicals and organophosphorus(有机磷) may covalently bind to the essential groups (e. g., —OH of Ser) in the active site of enzymes, resulting in irreversible inhibitions. These reactions, which may be called suicide(自杀) substrates, follow exponential decay(指数衰减) functions and are usually saturable(饱和的).

5.4　Regulatory enzymes 调节酶

5.4.1　Allosteric enzymes 别构酶

Allosteric proteins are those having "other shapes" or conformations(构象) induced by the binding of modulators(调节物). The same concept applies to certain regulatory enzymes, as conformational changes induced by one or more modulators interconvert(相互转化) more-active(活性较高) and less-active(活性较低) forms of the enzyme. The modulators for allosteric enzymes may be inhibitory or stimulatory(刺激的). Often the modulator is the substrate(底物) itself; regulatory enzymes for which substrate and modulator are identical are called homotropic enzymes(亲同酶). When the modulator is a molecule other than the substrate, the enzyme is said to be heterotropic enzyme(亲异酶). Note that allosteric modulators(别构调节物) should not be confused with noncompetitive or uncompetitive inhibitors. Although the latter bind at a second site on the enzyme, they do not necessarily mediate conformational changes(构象改变) between active and inactive forms, and the kinetic effects are distinct(不同).

The properties of allosteric enzymes are significantly different from those of simple non-regulatory enzymes. Some of the differences are structural. In addition to active sites, allosteric enzymes generally have one or more regulatory, or allosteric, sites for binding the modulator. Just as an enzyme's active site is specific for its substrate, each regulatory site is specific for its modulator. Enzymes with several modulators generally have different specific binding sites for each. In homotropic enzymes(亲同酶), the active site and regulatory site are the same.

5.4.2　Covalent modification　共价修饰

In another important class of regulatory enzymes, activity is modulated by covalent modification of the enzyme molecule. Modifying groups include phosphoryl, adenylyl, uridylyl, methyl, and adenosine diphosphate ribosyl groups. These groups are generally linked to and removed from the regulatory enzyme by separate enzymes.

修饰基团包括磷酰基,腺苷酸,尿苷酸,甲基和二磷酸腺苷核糖基团。

Phosphorylation(磷酸化) is the most common type of regulatory modification; one-third to one-half of all proteins in a eukaryotic(真核的) cell are phosphorylated. Some proteins have only one phosphorylated residue(残基), others have several, and a few have dozens of sites for phosphorylation. This mode of covalent modification is central to a large number of regulatory pathways. Phosphorylated enzymes may be more or less active than the dephosphorylated(去磷酸化) enzymes. Phosphorylated glycogen phosphorylase(糖原磷酸化酶), for example, is more active than dephosphorylated one, whereas the phosphorylated glycogen synthase(糖原合酶) is less active than the dephosphorylated enzyme.

磷酸化修饰是最常见的共价修饰方式。

举例说明通过磷酸化和去磷酸化的方式改变酶的活性。

5.4.3　Zymogen activation　酶原激活

For some enzymes, an inactive precursor(失活的前体) called a zymogen is cleaved(剪切) to form the active enzyme. Many proteolytic enzymes (proteases)(蛋白水解酶) of the stomach and pancreas are regulated in this way. Specific cleavage causes conformational changes that expose the enzyme active site. For example, pepsinogen(胃蛋白酶原) is secreted from the chief cells(主细胞) of the stomach and subsequently converted to active pepsin(胃蛋白酶) by the proteolytic removal

酶原失活的特点。

以胃蛋白酶原为例,解释酶原激活的过程。

微课：酶的结构与功能的关系

of a 44-amino acid peptide from its amino-terminus(氨基端). This reaction takes place at pH values below 5.0. Because this type of activation is irreversible(不可逆), other mechanisms are needed to inactivate these enzymes. Proteases(蛋白酶) are inactivated by inhibitor proteins that bind very tightly to the enzyme active site.

5.5　Enzyme Classification 酶的分类

酶命名方式举例。

　　Many enzymes have been named by adding the suffix(后缀) "-ase" to the name of their substrate or to a word or phrase describing their activity. Thus urease(尿素酶) catalyzes hydrolysis of urea(尿素), and DNA polymerase(聚合酶) catalyzes the polymerization(聚合) of nucleotides(核苷酸) to form DNA. Other enzymes were named by their discovers for a broad function, before the specific reaction catalyzed was known. For example, an enzyme known to act in the digestion of foods was named pepsin(胃蛋白酶), from the Greek *pepsis*, "digestion", and lysozyme(溶菌酶) was named for its ability to lyse(裂解) bacterial cell walls(细胞壁). Still others were named for their source(来源): trypsin(胰蛋白酶), named in part from the Greek *tryein*, "to wear down" was obtained by rubbing(擦) pancreatic tissue(胰腺组织) with glycerin(甘油). Sometimes the same enzyme has two or more names, or two different enzymes have the same name. Because of such ambiguities(模棱两可), and the ever-increasing number of newly discovered enzymes, biochemists, by international agreement, have adopted a system for naming(命名) and classifying(分类) enzymes. This system divides enzymes into six classes, each with sub-classes, based on the type of reaction catalyzed (Table below). Each enzyme is assigned a four-part classification number and a systematic(系统的) name, which identifies the reaction it catalyzes.

Table　International classification of enzymes.

No.	Class	Type of reaction catalyzed
1	Oxidoreductases	Transfer of electrons (hydride ions or H atoms)
2	Transferases	Group transfer reactions
3	Hydrolases	Hydrolysis reactions (transfer of functional groups to water)
4	Lyases	Addition of groups to double bonds, or formation of double bonds by removal of groups
5	Isomerases	Transfer of groups within molecules to yield isomeric forms
6	Ligases	Formation of C—C, C—S, C—O, and C—N bonds by condensation reactions coupled to ATP cleavage

根据酶催化反应的类型，国际上将酶分成六大类：
1. 氧化还原酶；
2. 转移酶；
3. 水解酶；
4. 裂合酶；
5. 异构酶；
6. 连接酶。

Note：Most enzymes catalyze the transfer of electrons(电子), atoms(原子), or functional groups(功能基团). They are therefore classified, given code numbers, and assigned names according to the type of transfer reaction(转移反应的类型), the group donor(基团供体), and the group acceptor(基团受体).

5.6　Enzyme application in industry 酶的工业应用

　　Enzymes are used in the chemical industry and other industrial applications when extremely specific catalysts are required. Enzymes in general are limited in the number of reactions they have evolved to catalyze and also by their lack of stability in organic solvents(有机溶剂) and at high temperatures. As a consequence, protein engineering is an active area of research and involves attempts to create new enzymes with novel properties, either through rational design (合理设计) or in vitro evolution. These efforts have begun to be successful, and a few enzymes have now been designed "from scratch" to catalyze reactions that do not occur in nature.

酶在生物洗涤剂、酿造业、烹饪、乳品业、食品加工以及淀粉行业中的应用举例。

Table Industrial applications of enzymes.

Application	Enzymes used	Uses
Biological detergent 生物洗涤剂	Proteases 蛋白酶 Amylases 淀粉酶 Lipases 脂肪酶	Remove protein, starch, and fat or oil stains from laundry and dishware.
Brewing industry 酿造业	Amylase 淀粉酶 Glucanases 葡聚糖酶 Proteases 蛋白酶 Amyloglucosidase 葡萄糖苷酶 Pullulanases 支链淀粉酶 Acetolactate decarboxylase 乙酰乳酸脱羧酶	Split polysaccharides and proteins in the malt. Make low-calorie beer and adjust ferment ability. Increase fermentation efficiency by reducing diacetyl formation.
Culinary uses 烹饪	Papain 木瓜蛋白酶	Tenderize meat for cooking.
Dairy industry 乳品业	Rennin 凝乳酶 Lipases 脂肪酶	Hydrolyze protein in the manufacture of cheese. Produce Camembert cheese and blue cheeses such as Roquefort.
Food processing 食品加工	Amylases 淀粉酶 Proteases 蛋白酶 Trypsin 胰蛋白酶 Cellulases 纤维素酶 Pectinases 果胶酶	Produce sugars from starch, such as in making high-fructose corn syrup. Lower the protein level of flour, as in biscuit-making. Manufacture hypoallergenic baby foods. Clarify fruit juices.
Starch industry 淀粉行业	Amylases 淀粉酶	Convert starch into glucose and various syrups.

【Key words】

enzyme 酶	K_M 米氏常数
substrate 底物	activation energy 活化能
active site 活性中心	competitive inhibition 竞争性抑制
binding site 结合部位	non-competitive inhibition 非竞争性抑制
catalytic site 催化部位	uncompetitive inhibition 反竞争性抑制
essential group 必需基团	zymogen activation 酶原激活
cofactor 辅因子	allosteric regulation 别构调节
coenzyme 辅酶	covalent modification 共价修饰
prosthetic group 辅基	allosteric enzyme 别构酶
specificity 专一性	isoenzyme 同工酶
Michaelis-Menten equation 米-曼氏方程	ribozyme 核酶

Questions

1. An enzyme facilitates chemical reaction by (　　)
   A. increasing the free-energy difference between reactants and products.
   B. decreasing the free-energy difference between reactants and products.
   C. lowering the activation energy of the reaction.
   D. raising the activation energy of the reaction.
2. The specificity of enzyme is determined by (　　)
   A. coenzyme　　　B. apoenzyme　　　C. prosthetic group　　　D. metal ion
3. Explain the terms of enzyme, active site, reversible/irreversible inhibition, isoenzyme, zymogen activation, and allosteric enzyme.
4. Both glucokinase and hexokinase can convert glucose to glucose-6-P. The former has a K_M of about 10 mmol/L, the later about 0.1 mmol/L. Which one catalyzes the production of glucose-6-P faster, why?
5. If the enzymatic velocity is 90% of V_{max}, [S] should be (　　) times of K_M.
   A. 4.5　　　　B. 9　　　　C. 8　　　　D. 5

References

[1] McKee Trudy, R McKee James. Biochemistry: An Introduction (Second Edition)[M]. New York: McGraw-Hill Companies, 1999.

[2] Nelson D L, Cox M M. Lehninger Principles of Biochemistry (Third Edition)[M]. Derbyshire: Worth Publishers, 2000.

[3] Ding W, Jia H T. Biochemistry (First Edition)[M]. Beijing: Higher Education Press, 2012.

[4] 郑集,陈钧辉.普通生物化学(第四版)[M].北京:高等教育出版社,2007.

[5] 王艳萍.生物化学[M].北京:中国轻工业出版社,2013.

Part II
Metabolism

新陈代谢

Chapter 1　Overview of Metabolism 代谢总论

　　The concepts of conformation(构象) and dynamics(动力学) — especially those dealing with the specificity(专一性) and catalytic power(催化能力) of enzymes, the regulation of their catalytic activity, and the transport of molecules and ions across membranes—enable us to now ask questions fundamental to biochemistry：

　　1. How does a cell extract energy(获取能量) and reducing power(还原能力) from its environment?

一个细胞如何从环境中获取能量。

　　2. How does a cell synthesize the building blocks of its macromolecules and then the macromolecules themselves?

一个细胞如何合成和构建大分子。

　　These processes are carried out by a highly integrated network of chemical reactions that are collectively known as metabolism(新陈代谢) or intermediary metabolism(中间代谢).

　　Metabolism is the set of chemical reactions that happen in living organisms(生物体) to maintain(维持) life. These processes allow organisms to grow and reproduce(繁殖), maintain their structures(结构), and respond to(应答) their environments. Metabolism is usually divided into (分为) two categories(类别). Catabolism(分解代谢) breaks down(分解) organic matter(有机物), for example to harvest (获得) energy in cellular respiration(细胞呼吸). Anabolism(合成代谢) uses energy to construct(构造) components of cells such as proteins(蛋白质) and nucleic acids(核酸).

新陈代谢维持生物体生命活动。

　　Enzymes are crucial(重要) to metabolism because they allow organisms to drive desirable reactions(驱动所需的反应) that require energy and will not occur by themselves, by coupling(耦合) them to spontaneous reactions(自发反应) that release(释放) energy. As enzymes act as catalysts(催化剂), they allow these reactions to proceed quickly and efficiently(高效地). Enzymes also allow the regulation(调节) of metabolic

酶在新陈代谢中的重要作用。

pathways in response to changes in the cell's environment or signals(信号) from other cells.

新陈代谢对生物体的作用。

The metabolism of an organism determines(确定) which substances(物质) it will find nutritious(营养) and which it will find poisonous(有毒的). For example, some prokaryotes(原核生物) use hydrogen sulfide(硫化氢) as a nutrient(养分), yet this gas is poisonous to animals. The speed of metabolism, the metabolic rate, also influences how much food an organism will require. A striking feature(显著特征) of metabolism is the similarity(相似) of the basic metabolic pathways and components between even vastly different species.

新陈代谢的显著特征是不同物种间在基本代谢途径和组成上的相似性。

1.1 Key biochemicals 重要的生化分子

许多重要物质,如氨基酸、碳水化合物和脂类分子等参与了代谢。

Most of the structures that make up animals, plants and microbes(微生物) are made from three basic classes of molecule: amino acids(氨基酸), carbohydrates(碳水化合物) and lipids(脂类)(often called fats). As these molecules are vital(重要) for life, metabolic reactions either focus on(集中于) making these molecules during the construction(构建) of cells and tissues(组织), or breaking them down and using them as a source of energy, in the digestion(消化) and use of food. Many important biochemicals can be joined together to make polymers(聚合物) such as DNA and proteins. These macromolecules are essential.

1.1.1 Amino acids and proteins 氨基酸和蛋白质

蛋白质的组成及作用。

Proteins are made of amino acids arranged(排列) in a linear chain(线性链) and joined together by peptide bonds(肽键). Many proteins are the enzymes that catalyze(催化) the chemical reactions in metabolism. Other proteins have structural or mechanical(机械的) functions, such as the proteins that form the cytoskeleton(骨架), a system of scaffolding(搭建系统) that maintains the cell shape. Proteins are also important in cell signaling(信号), immune responses(免疫反应), cell adhesion(细胞黏附), active transport across membranes(跨膜运输), and the cell cycle(细胞周期).

1.1.2　Lipids 脂质

Lipids(脂类) are the most diverse(最多样化的) group of biochemicals. Their main structural uses are as part of biological membranes(细胞膜) such as the cell membrane, or as a source of energy. Lipids are usually defined(定义) as hydrophobic(疏水) or amphipathic(两性) biological molecules that will dissolve(溶解于) in organic solvents(有机溶剂) such as benzene(苯) or chloroform(氯仿). The fats are a large group of compounds that contain fatty acids(脂肪酸) and glycerol(甘油); a glycerol molecule attached to three fatty acid esters(脂肪酸酯) is a triacylglyceride(三酰甘油). Several variations(变异) on this basic structure exist, including alternate backbones such as sphingosine(鞘氨醇) in the sphingolipids(鞘脂类), and hydrophilic groups(亲水基团) such as phosphate(磷酸) in phospholipids(磷脂). Steroids(类固醇) such as cholesterol(胆固醇) are another major class of lipids that are made in cells.

脂类的基本特点。

Key biochemicals.

Type of molecule	Name of monomer forms	Name of polymer forms	Examples of polymer forms
Amino acids	Amino acids	Proteins(also called polypeptides)	Fibrous proteins and globular proteins
Carbohydrates	Monosaccharides	Polysaccharides	Starch, glycogen and cellulose
Nucleic acids	Nucleotides	Polynucleotides	DNA and RNA

1.1.3　Carbohydrates 糖类

Carbohydrates(碳水化合物) are straight-chain(直链) aldehydes(醛) or ketones(酮) with many hydroxyl groups(羟基) that can exist as straight chains(链状) or rings(环状).

碳水化合物的基本特点。

Chain carbohydrates are the most abundant biological molecules, and fill numerous(众多) roles, such as the storage

（储存）and transport（运输）of energy(starch 淀粉，glycogen 糖原）and structural components (cellulose in plants 植物纤维素，chitin in animals). The basic carbohydrate units are called monosaccharides（单糖）and include galactose（半乳糖），fructose（果糖），and most importantly glucose（葡萄糖）. Monosaccharides can be linked together to form polysaccharides（多糖）in almost limitless（无限）ways.

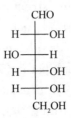

Glucose can exist in both a straight-chain.

1.1.4 Nucleotides 核苷酸

核酸与核苷酸的基本特点。

The two nucleic acids（核酸），DNA and RNA are polymers（聚合物）of nucleotides（核苷酸），each nucleotide comprising a phosphate group（磷酸基），a ribose sugar group（核糖组），and a nitrogenous base（含氮碱基）. Nucleic acids are critical（关键的）for the storage（存储）and use of genetic information（遗传信息），through the processes of transcription（转录）and protein biosynthesis（蛋白质生物合成）. This information is protected by DNA repair mechanisms（DNA修复机制）and propagated（传播）through DNA replication（复制）. Many viruses（病毒）have an RNA genome（基因组），for example HIV（艾滋病毒），which uses reverse transcription（逆转录）to create a DNA template（模板）from its viral RNA genome. RNA in ribozymes（核酶）such as spliceosomes（剪接体）and ribosomes（核糖体）is similar to enzymes as it can catalyze（催化）chemical reactions. Individual（单个）nucleosides are made by attaching a nucleobase（碱基）to a ribose sugar. These bases are heterocyclic rings containing nitrogen（含氮杂环），classified as purines or pyrimidines（嘌呤或嘧啶）. Nucleotides also act as coenzymes（辅酶）in metabolic group transfer reactions.

1.1.5 Coenzymes 辅酶

Metabolism involves a vast array of chemical reactions, but most fall under a few basic types of reactions that involve the transfer of functional groups(官能团的转移). This common chemistry allows cells to use a small set of metabolic intermediates(代谢中间体) to carry chemical groups between different reactions. These group-transfer intermediates(转移中间体) are called coenzymes. Each class of group-transfer reaction is carried out by a particular(特定的) coenzyme, which is the substrate(底物) for a set of enzymes that produce it, and a set of enzymes that consume(消耗) it. These coenzymes are therefore continuously being made, consumed and then recycled(循环利用). One central(中心的) coenzyme is adenosine triphosphate（ATP, 三磷酸腺苷）, the universal energy currency(现金) of cells. This nucleotide(核苷酸) is used to transfer chemical energy between different chemical reactions. There is only a small amount of ATP in cells, but as it is continuously regenerated(再生), the human body can use about its own weight in ATP per day. ATP acts as a bridge(桥) between catabolism and anabolism(分解代谢和合成代谢), with catabolic reactions generating ATP and anabolic reactions consuming it. It also serves as a carrier(载体) of phosphate groups in phosphorylation reactions(磷酸化反应).

A vitamin(维生素) is an organic compound needed in small quantities that cannot be made in the cells. In human nutrition (营养), most vitamins function as coenzymes after modification (修改), for example, all water-soluble vitamins(水溶性维生素) are phosphorylated(磷酸化) or are coupled to nucleotides(耦合核苷酸) when they are used in cells. Nicotinamide adenine dinucleotide(NADH,烟酰胺腺嘌呤二核苷酸), a derivative(衍生物) of vitamin B$_3$（niacin, 烟酸）, is an important coenzyme that acts as a hydrogen acceptor(受体). Hundreds of separate (单独的) types of dehydrogenases(脱氢酶) remove electrons from their substrates and reduce NAD$^+$ into NADH. This reduced form of the coenzyme is then a substrate for any of the reductases(还原酶) in the cell that need to reduce their substrates. Nicotinamide adenine dinucleotide exists in two related forms in the cell, NADH and NADPH. The NAD^+/NADH form is more important in catabolic reactions, while $NADP^+$/NADPH is used in anabolic reactions.

1.1.6 Minerals and cofactors 金属和辅助因子

无机元素有重要的作用。

Inorganic elements(无机元素) play critical(重要的) roles in metabolism; some are abundant (e. g., Sodium and potassium, 例如钠和钾) while others function at minute(分) concentrations(浓度). About 99% of mammals' mass are the elements(大约99%的哺乳动物的质量元素) carbon, nitrogen (氮), calcium(钙), sodium(钠), chlorine(氯), potassium(钾), hydrogen(氢), phosphorus(磷), oxygen(氧) and sulfur(硫). The organic compounds (proteins, lipids and carbohydrates) contain the majority of the carbon and nitrogen and most of the oxygen and hydrogen is present as water.

无机元素作为离子电解质。

The abundant inorganic elements(无机元素) act as ionic electrolytes(离子电解质). The most important ions(离子) are sodium(钠), potassium(钾), calcium(钙), magnesium(镁), chloride(氯), phosphate(磷酸), and the organic ion bicarbonate (有机碳酸氢离子). The maintenance of precise gradients across cell membranes(维持准确的跨细梯度) maintains osmotic pressure and pH(保持渗透压梯度和pH). Ions are also critical (至关重要的) for nerves(神经) and muscles(肌肉), as action potentials(电位) in these tissues(组织) are produced by the exchange of electrolytes between the extracellular fluid(细胞外液) and the cytosol(细胞质). Electrolytes(电解质) enter and leave cells through proteins in the cell membrane called ion channels(离子通道). For example, muscle contraction(肌肉收缩) depends upon(取决于) the movement of calcium(钙的运动), sodium(钠) and potassium(钾) through ion channels(离子通道) in the cell membrane(细胞膜) and T-tubules(小管).

The transition metals(过渡金属) are usually present as trace elements(微量元素) in organisms, with zinc(锌) and iron (铁) being most abundant. These metals are used in some proteins as cofactors(辅助因子) and are essential for the activity of enzymes(酶的活性) such as catalase(过氧化氢酶) and oxygen-carrier proteins(运输氧的蛋白质) such as hemoglobin(血红蛋白). These cofactors are bound tightly(紧密地) to a specific protein(特定的蛋白质); although enzyme cofactors can be modified during catalysis(催化), cofactors always return to their original state(原始状态) after catalysis has taken place(发生). The metal micronutrients(金属微量元素) are taken up into organisms by specific transporters(特定的转运) and bound to storage(贮藏) proteins such as ferritin or metallothionein(铁蛋白或金属硫蛋白) when not being used.

金属微量元素可作为酶的辅助因子。

1.2 Catabolism 分解代谢

Catabolism is the set of metabolic processes that break down large molecules(分解大分子). These include breaking down and oxidizing(氧化) food molecules. The purpose(目的) of the catabolic reactions is to provide the energy and components needed by anabolic reactions(合成代谢反应). The exact nature(实质) of these catabolic reactions differ from organism to organism and organisms can be classified based on(基于) their sources of energy and carbon (their primary nutritional groups), as shown in the table below(如下表所示). Organic molecules(有机分子) being used as a source of energy in organotrophs(有机营养菌), while lithotrophs(无机生物) use inorganic substrates(无机基质) and phototrophs capture sunlight(光养捕获阳光) as chemical energy. However, all these different forms of metabolism depend on redox reactions(氧化还原反应) that involve the transfer of electrons from reduced donor molecules(供体分子) such as organic molecules, water, ammonia(氨), hydrogen sulfide(硫化氢) or ferrous ions (亚铁离子) to acceptor molecules(受体分子) such as oxygen, nitrate(硝酸) or sulfate(硫酸). In animals these reactions involve complex organic molecules being broken down to simpler molecules, such as carbon dioxide(二氧化碳) and water. In photosynthetic organisms(光合生物) such as plants

分解代谢的过程。

and cyanobacteria(蓝细菌), these electron-transfer reactions(电子转换反应) do not release(释放) energy, but are used as a way of storing energy absorbed from sunlight.

Classification of organisms based on their metabolism.

energy source	sunlight	photo-		
	preformed molecules	chemo-		
electron donor	organic compound		organo-	-troph
	inorganic compound		litho-	
carbon source	organic compound		hetero-	
	inoyganic compound		auto-	

动物中常见的分解反应分为三个主要阶段。

The most common set of catabolic reactions in animals can be separated into three main stages(主要阶段). In the first, large organic molecules such as proteins, polysaccharides or lipids are digested(消化) into their smaller components outside cells. Next, these smaller molecules are taken up by cells and converted(转化) to yet smaller molecules, usually acetyl coenzyme A (acetyl-CoA)(乙酰辅酶A), which releases some energy. Finally, the acetyl group(乙酰基) on the CoA is oxidized(氧化) to water and carbon dioxide(二氧化碳) in the citric acid cycle(三羧酸循环) and electron transport chain(电子传递链), releasing the energy that is stored by reducing the coenzyme nicotinamide adenine dinucleotide(烟酰胺腺嘌呤二核苷酸)(NAD^+) into NADH.

1.2.1 Digestion 消化

大分子通过消化酶分解成更小的单位才可以用于细胞代谢。

Macromolecules(大分子) such as starch(淀粉), cellulose(纤维素) or proteins cannot be rapidly(迅速) taken up by cells and need to be broken into their smaller units before they can be used in cell metabolism. Several common classes of enzymes digest these polymers. These digestive enzymes(消化酶) include proteases(蛋白酶) that digest proteins into amino acids(氨基酸), as well as glycoside hydrolases(糖苷水解酶) that digest polysaccharides into monosaccharides.

Microbes simply secrete digestive enzymes(消化酶) into their surroundings(周围环境), while animals only secrete these enzymes from specialized cells in their guts. The amino acids or sugars released by these extracellular enzymes(胞外酶) are then pumped into cells by specific active transport proteins(特定的主动转运蛋白).

动物只能由特殊细胞分泌消化酶。

1.2.2 Energy from organic compounds 能量来自有机化合物

Carbohydrate catabolism is the breakdown of carbohydrates into smaller units. Carbohydrates are usually taken into cells once they have been digested into monosaccharides. Once inside, the major route of breakdown is glycolysis, where sugars such as glucose and fructose are converted into pyruvate(丙酮酸) and some ATP is generated. Pyruvate is an intermediate in several metabolic pathways, but the majority is converted to acetyl-CoA and fed into the citric acid cycle(三羧酸循环). Although some more ATP is generated in the citric acid cycle, the most important product is NADH, which is made from NAD^+ as the acetyl-CoA is oxidized. This oxidation releases carbon dioxide as a waste product. In anaerobic conditions(在厌氧条件下), glycolysis produces lactate(乳酸), through the enzyme lactate dehydrogenase(乳酸脱氢酶) re-oxidizing NADH to NAD^+ for re-use in glycolysis. An alternative route for glucose breakdown is the pentose phosphate pathway(磷酸戊糖途径), which reduces the coenzyme NADPH and produces pentose sugars such as ribose, the sugar component of nucleic acids.

丙酮酸是重要的中间产物。

Fats are catabolized by hydrolysis(水解) to free fatty acids (游离脂肪酸) and glycerol(甘油). The glycerol enters glycolysis and the fatty acids are broken down by beta oxidation to release(释放) acetyl-CoA(乙酰辅酶 A), which then is fed into the citric acid cycle (三羧酸循环). Fatty acids release more energy upon oxidation than carbohydrates because carbohydrates contain more oxygen(氧气) in their structures.

脂肪水解产生游离脂肪酸与甘油。

脂肪酸氧化释放的能量比糖类更高。

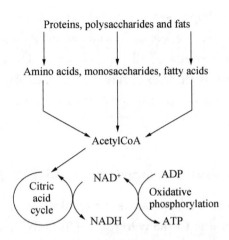

A simplified outline of the catabolism of proteins, carbohydrates and fats.

Amino acids(氨基酸) are either used to synthesize(合成) proteins and other biomolecules, or oxidized to urea(尿素) and carbon dioxide as a source of energy. The oxidation pathway starts with the removal of the amino group by a transaminase (移除转氨酶). The amino group(氨基) is fed into the urea cycle(尿素循环), leaving a deaminated carbon skeleton(碳架) in the form of a keto acid(酮酸). Several of these keto acids(酮酸) are intermediates(中间体) in the citric acid cycle, for example, the deamination of glutamate forms(谷氨酸脱氨基作用) α-ketoglutarate(α-酮戊二酸). The glucogenic amino acids can also be converted into glucose, through gluconeogenesis(糖异生) (discussed below).

氨基酸的代谢。

1.3　Energy transformations 能量转换

1.3.1　Oxidative phosphorylation 氧化磷酸化

In oxidative phosphorylation, the electrons(电子) removed from organic molecules(有机分子) in areas such as the protagon acid cycle(初磷脂酸循环) are transferred to oxygen and the energy released is used to make ATP. This is done in eukaryotes(真核生物) by a series of proteins in the membranes of mitochondria(线粒体膜) called the electron transport chain

氧化磷酸化中的电子转移。

（电子传输链）. In prokaryotes（原核生物）, these proteins are found in the cell's inner membrane（细胞内膜）. These proteins use the energy released from passing electrons from reduced molecules like NADH onto oxygen to pump protons across a membrane（跨膜泵出质子）.

Pumping protons out of the mitochondria（线粒体）creates a proton concentration difference across the membrane（跨膜质子浓度差）and generates an electrochemical gradient（形成一个电化学梯度）. This force drives protons back into the mitochondrion through the base of an enzyme called ATP synthase（ATP合酶）. The flow of protons（质子流）makes the stalk subunit（杆单元）rotate（旋转）, causing the active site of the synthase domain（结构域）to change shape and phosphorylate adenosine diphosphate（磷酸化腺苷二磷酸）- turning it into ATP.

质子进入线粒体造成浓度差和电化学梯度，生成ATP。

1.3.2　Energy from inorganic compounds 能量来自无机化合物

Chemolithotrophy（无机化能营养）is a type of metabolism found in prokaryotes（原核生物）where energy is obtained from the oxidation of inorganic compounds. These organisms can use hydrogen（氢气）, reduced sulfur compounds（硫化合物）(such as sulfide, hydrogen sulfide and thiosulfate,如硫、硫化氢和硫代硫酸盐), ferrous iron（二价铁）(FeII) or ammonia（氨）as sources of reducing power and they gain energy from the oxidation of these compounds with electron acceptors（电子受体）such as oxygen or nitrite（亚硝酸盐）. These microbial processes are important in global biogeochemical cycles such as acetogenesis（乙酸化）, nitrification（硝化）and denitrification（反硝化）and are critical for soil fertility（土壤肥力）.

无机化合物的氧化。

1.4　Anabolism 合成代谢

Anabolism is the set of constructive（构造性的）metabolic

合成代谢的基本性质。

processes where the energy released by catabolism is used to synthesize complex molecules. In general, the complex molecules that make up cellular structures are constructed step-by-step(逐步) from small and simple precursors(前体). Anabolism involves three basic stages. Firstly, the production of precursors such as amino acids, monosaccharides, isoprenoids(类异戊二烯) and nucleotides(核苷酸), secondly, their activation into reactive forms using energy from ATP, and thirdly, the assembly of these precursors into complex molecules such as proteins, polysaccharides, lipids and nucleic acids.

自养生物和异养生物的分子细胞构造与能量来源。

Organisms differ in how many of the molecules in their cells they can construct for themselves. Autotrophs(自养生物) such as plants can construct the complex organic molecules in cells such as polysaccharides and proteins from simple molecules like carbon dioxide and water. Heterotrophs(异养生物), on the other hand(另一方面), require a source of more complex substances, such as monosaccharides and amino acids, to produce these complex molecules. Organisms can be further classified by ultimate(最终) source of their energy: photoautotrophs(光合自养生物) and photoheterotrophs(光异养生物) obtain energy from light, whereas chemoautotrophs(化能自养生物) and chemoheterotrophs(化能异养生物) obtain energy from inorganic oxidation reactions(氧化反应).

蓝细菌和藻类植物的光合作用。

Photosynthesis(光合作用) is the synthesis of carbohydrates from sunlight and carbon dioxide(CO_2). In plants, cyanobacteria and algae(蓝细菌和藻类植物), oxygenic photosynthesis splits water, with oxygen produced as a waste product. This process uses the ATP and NADPH produced by the photosynthetic reaction centers, as described above, to convert CO_2 into glycerate 3-phosphate(甘油酸三酯), which can then be converted into glucose. This carbon-fixation reaction is carried out by the enzyme RuBisCO(二磷酸核酮糖羧化酶) as part of the Calvin-Benson cycle. Three types of photosynthesis occur in plants, C3 carbon fixation, C4 carbon fixation and CAM(景天酸代谢) photosynthesis. These differ by the route that carbon dioxide takes to the Calvin cycle(卡尔文循环), with C3 plants fixing CO_2 directly, while C4 and CAM photosynthesis incorporate the CO_2 into other compounds first, as adaptations to deal with intense sunlight and dry conditions.

In photosynthetic prokaryotes the mechanisms of carbon fixation are more diverse. Here, carbon dioxide can be fixed by the Calvin-Benson cycle, a reversed citric acid cycle(三羧酸循环), or the carboxylation of acetyl-CoA(乙酰辅酶A的羧化作用). Prokaryotic chemoautotrophs also fix CO_2 through the Calvin-Benson cycle, but use energy from inorganic compounds to drive the reaction.

光合原核生物固碳作用的机制多样化。

1.4.1 Carbohydrates and glycans 糖类与聚糖

In carbohydrate anabolism(合成代谢), simple organic acids(有机酸) can be converted into monosaccharides such as glucose and then used to assemble polysaccharides such as starch(淀粉). The generation of glucose from compounds like pyruvate(丙酮酸), lactate(乳酸), glycerol(甘油), glycerate 3-phosphate(3-磷酸甘油酸) and amino acids is called gluconeogenesis. Gluconeogenesis converts pyruvate to glucose-6-phosphate(6-磷酸葡萄糖) through a series of intermediates(中间体), many of which are shared with glycolysis(糖酵解). However, this pathway is not simply glycolysis run in reverse(逆向), as several steps are catalyzed by non-glycolytic enzymes. This is important as it allows the formation and breakdown of glucose to be regulated separately and prevents both pathways from running simultaneously(同时地) in a futile cycle(无效循环).

碳水化合物的分解代谢。

Although fat is a common way of storing energy, in vertebrates(脊椎动物) such as humans the fatty acids in these stores cannot be converted to glucose through gluconeogenesis(糖异生) as these organisms cannot convert acetyl-CoA into pyruvate; plants do, but animals do not, have the necessary enzymatic machinery(酶学机制). As a result, after long-term starvation(长期饥饿), vertebrates need to produce ketone bodies(酮体) from fatty acids to replace glucose in tissues(组织) such as the brain that cannot metabolize fatty acids. In other organisms such as plants and bacteria, this metabolic problem is solved using the glyoxylate cycle（乙醛酸循环）, which by passes(绕过) the decarboxylation(脱羧反应) step in the citric acid cycle and allows the transformation(转化) of acetyl-CoA to oxaloacetate(草酰乙酸), where it can be used for the production of glucose.

乙醛酸循环。

Polysaccharides and glycans（聚糖）are made by the sequential addition of monosaccharides（单糖）by glycosyltransferase（糖基转移酶）from a reactive sugar-phosphate（活性磷酸糖）donor（供者）such as uridine diphosphate glucose（UDP-glucose）（尿苷二磷酸葡萄糖）to an acceptor hydroxyl group（受体羟基）on the growing polysaccharide. As any of the hydroxyl groups on the ring of the substrate(底物分子环上的任意羟基) can be acceptors(受体), the polysaccharides produced can have straight(直的) or branched(分支的) structures. The polysaccharides produced can have structural or metabolic functions themselves, or be transferred to lipids and proteins by enzymes called oligosaccharyltransferases（寡糖转移酶）.

多糖和聚糖的合成代谢。

1.4.2 Fatty acids, isoprenoids and steroids 脂肪酸,异戊二烯和类固醇

Fatty acids are made by fatty acid synthases(脂肪酸合酶) that polymerize(聚合) and then reduce(还原) acetyl-CoA units. The acyl chains(酰基链) in the fatty acids are extended(延长) by a cycle of reactions that add the actyl group, reduce it to an alcohol（乙醇）,dehydrate(脱水) it to an alkene group(烯烃基团) and then reduce it again to an alkane group(烷烃基团). The enzymes of fatty acid biosynthesis are divided into two groups, in animals and fungi（真菌）all these fatty acid synthase reactions are carried out by a single multifunctional type I protein(多功能Ⅰ型蛋白质), while in plant plastids(叶绿体)

脂肪酸合成。

动物和真菌中的脂肪酸合成酶与植物或细菌中的酶不同。

and bacteria separate（不同的）type II enzymes（II 型酶）perform each step in the pathway.

Terpenes（萜烯）and isoprenoids（类异戊二烯）are a large class of lipids that include the carotenoids（类胡萝卜素）and form the largest class of plant natural products. These compounds are made by the assembly and modification of isoprene（异戊二烯）units donated from the reactive precursors（活性前体）isopentenyl pyrophosphate（异戊烯焦磷酸）and dimethylallyl pyrophosphate（二甲基烯丙基焦磷酸）. These precursors can be made in different ways. In animals and archaea（古细菌）, the mevalonate（甲羟戊酸）pathway produces these compounds from acetyl-CoA, while in plants and bacteria the non-mevalonate pathway uses pyruvate and glyceraldehyde 3-phosphate（甘油醛-3-磷酸）as substrates. One important reaction that uses these activated isoprene donors is steroid（类固醇）biosynthesis. Here, the isoprene units are joined together to make squalene（角鲨烯）and then folded up（折叠起来）and formed into a set of rings to make lanosterol（羊毛甾醇）. Lanosterol can then be converted into other steroids such as cholesterol（胆固醇）and ergosterol（麦角固醇）.

萜烯和类异戊二烯前体物质通过不同的方式生成。

1.4.3 Proteins 蛋白质

Organisms vary in their ability to synthesize the 20 common amino acids（20 种常见氨基酸）. Most bacteria and plants can synthesize all twenty, but mammals（哺乳动物）can synthesize only eleven nonessential amino acids（非必需氨基酸）. Thus, nine essential amino acids（9 种必需氨基酸）must be obtained from food. All amino acids are synthesized from intermediates in glycolysis（糖酵解）, the citric acid cycle（三羧酸循环）, or the pentose phosphate pathway（磷酸戊糖途径）. Nitrogen（氮）is provided by glutamate（谷氨酸）and glutamine（谷氨酰胺）. Amino acid synthesis depends on the formation of the appropriate（适当的）alpha-keto acid（酮酸）, which is then transaminated（促转氨）to form an amino acid.

氨基酸合成。

蛋白质的生物合成。

Amino acids are made into proteins by being joined together in a chain by peptide bonds(肽键). Each different protein has a unique sequence of amino acid residues(特定的氨基酸残基序列): this is its primary structure. Just as the letters of the alphabet(字母表) can be combined to form an almost endless(无尽的) variety of words, amino acids can be linked in varying sequences to form a huge(巨大的) variety of proteins. Proteins are made from amino acids that have been activated by attachment(配基) to a transfer RNA molecule through an ester bond(酯键). This aminoacyl-tRNA precursor(氨酰基前体-tRNA) is produced in an ATP-dependent reactioncarried out by an aminoacyl tRNA synthetase(氨酰 tRNA 合成酶). This aminoacyl-tRNA is then a substrate for the ribosome(核糖体), which joins the amino acid onto the elongating protein chain(延伸蛋白质链), using the sequence information in a messenger RNA.

1.4.5 Nucleotide synthesis and salvage 核苷酸合成与补救

核苷酸补救合成途径更高效。嘌呤合成方式。

Nucleotides are made from amino acids, carbon dioxide(二氧化碳) and formic acid(甲酸) in pathways that require large amounts of metabolic energy. Consequently(因此), most organisms have efficient(高效的) systems(系统) to salvage(救助) preformed(预制) nucleotides. Purines are synthesized as nucleosides(bases 碱基 attached to ribose 核糖). Both adenine(腺嘌呤) and guanine(鸟嘌呤) are made from the precursor nucleoside inosine monophosphate(次黄苷酸的核苷前体), which is synthesized using atoms from the amino acids glycine(甘氨酸), glutamine(谷氨酰胺), and aspartic acid(天冬氨酸), as well as formate(甲酸盐) transferred from the coenzyme(辅酶) tetrahydrofolate(四氢叶酸). Pyrimidines, on the other hand, are synthesized from the base orotate(乳清酸), which is formed from glutamine(谷氨酰胺) and aspartate(天冬氨酸).

1.5 Thermodynamics of living organisms 生物热力学

Living organisms must obey the laws of thermodynamics（活体生物必须遵循热力学定律）, which describe the transfer of heat(热量) and work. The second law of thermodynamic (热力学第二定律) states (指出) that in any closed system(在任何封闭的体系中), the amount of entropy (disorder) will tend to increase（熵（无序程度）的量会增加）. Although living organisms' amazing complexity(复杂性) appears to contradict this law （似乎与本定律抵触的）, life is possible as all organisms are open systems that exchange matter and energy with their surroundings. Thus living systems are not in equilibrium(平衡), but instead are dissipative systems(耗散系统) that maintain their state of high complexity by causing a larger increase in the entropy of their environments. The metabolism of a cell achieves this by coupling the spontaneous processes of catabolism(分解代谢的自发过程) to the non-spontaneous processes of anabolism(分解代谢的非自发过程). In thermodynamic terms(在热力学方面), metabolism maintains order by creating disorder(形成无序).

生物体代谢遵循热力学定律。

1.6 Regulation and control 调节与控制

As the environments of most organisms are constantly changing(不断变化的), the reactions of metabolism must be finely regulated to maintain a constant set of conditions（一个恒定的环境）within cells, a condition called homeostasis（体内平衡）. Metabolic regulation also allows organisms to respond to signals（响应信号）and interact actively(积极响应) with their environments. Two closely linked concepts(两个密切相关的概念) are important for understanding how metabolic pathways are controlled. Firstly, the regulation of an enzyme in a pathway is how its activity is increased and decreased in response to signals. Secondly, the control exerted by this

大多数生物必须精确调节新陈代谢以适应环境的不断变化。

enzyme is the effect that these changes in its activity have on the overall rate(总速率) of the pathway (the flux through the pathway)(途径的通量). For example, an enzyme may show large changes in activity (i.e. it is highly regulated) but if these changes have little effect on the flux of a metabolic pathway, then this enzyme is not involved in the control of the pathway.

There are multiple(多种的) levels of metabolic regulation. In intrinsic regulation(内在调节), the metabolic pathway self-regulates to respond to changes in the levels of substrates or products, for example, a decrease in the amount of product can increase the flux through the pathway to compensate(补偿). This type of regulation often involves allosteric regulation(变构调节) of the activities of multiple enzymes in the pathway. Extrinsic control(外在控制) involves a cell in a multicellular organism(多细胞生物) changing its metabolism in response to signals from other cells. These signals are usually in the form of soluble(可溶性) messengers(信使) such as hormones(激素) and growth factors(生长因素) and are detected(检测) by specific receptors(特定的受体) on the cell surface. These signals are then transmitted(传送) inside the cell by second messenger systems that often involved the phosphorylation(磷酸化) of proteins.

生物体内有多种代谢途径控制。

A very well understood example of extrinsic control(外在控制) is the regulation of glucose metabolism by the hormone insulin(胰岛素). Insulin is produced in response to rises in blood glucose levels(血糖水平). Binding of the hormone to insulin receptors on cells then activates a cascade of protein kinases(激活蛋白激酶级联) that cause the cells to take up glucose and convert it into(转换为) storage molecules such as fatty acids and glycogen. The metabolism of glycogen is controlled by activity of phosphorylase(磷酸化酶), the enzyme that breaks down(分解) glycogen, and glycogen synthase, the enzyme that makes it. These enzymes are regulated in a reciprocal fashion(互惠的方式), with phosphorylation inhibiting glycogen synthase, but activating phosphorylase. Insulin causes glycogen synthesis by activating protein phosphatases and producing a decrease in the phosphorylation of these enzymes.

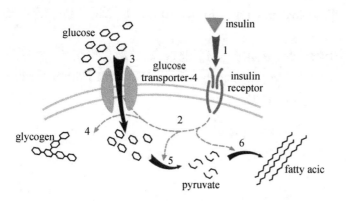

胰岛素对葡萄糖摄取与代谢的调控作用。

Effect of insulin on glucose uptake and metabolism. Insulin binds to its receptor (1) which in turn starts many protein activation cascades (2). These include: translocation of Glut-4 transporter to the plasma membrane and influx of glucose (3), glycogen synthesis (4), glycolysis (5) and fatty acid synthesis (6).

【Key words】

metabolism 代谢	oxidative phosphorylation 氧化磷酸化
catabolism 分解代谢	insulin 胰岛素
anabolism 合成代谢	salvage 补救
photosynthesis 光合作用	dynamics 动力学
mitochondria 线粒体	thermodynamics 热力学
biochemicals 生化分子	reduce 还原
glycans 聚糖	nicotinamide adenine dinucleotide(NADH) 烟酰胺腺嘌呤二核苷酸

Questions

1. What is meant by the intermediary metabolism?
2. Differention between anabolism and catabolism.
3. Why does it make good sense to have a single nucleotide, ATP, function as the cellular energy currency?

References

[1] McKee Trudy, R McKee James. Biochemistry: An Introduction (Second Edition)[M]. New York: McGraw-Hill Companies, 1999.

[2] Nelson D L, Cox M M. Lehninger Principles of Biochemistry (Third Edition)[M]. Derbyshire: Worth Publishers, 2000.

[3] Ding W, Jia H T. Biochemistry (First Edition)[M]. Beijing: Higher Education Press, 2012.

[4] 郑集,陈钧辉. 普通生物化学(第四版)[M]. 北京:高等教育出版社,2007.

[5] Berg J M, Tymoczko J L, Styer L. Biochemistry (Seventh Edition)[M]. New York: W. H. Freeman and Company, 2012.

Chapter 2　Carbohydrate metabolism 糖代谢

2.1　Glycolysis 糖酵解

Glycolysis(糖酵解)(from glycose(葡萄糖), an older term for glucose + -lysis degradation) is the metabolic pathway(代谢途径) that converts glucose $C_6H_{12}O_6$, into pyruvate(丙酮酸), $CH_3COCOO^- + H^+$. The free energy released in this process is used to form the high-energy compounds(高能化合物) ATP(adenosine triphosphate, 三磷酸腺苷) and NADH(reduced nicotinamide adenine dinucleotide, 烟酰胺腺嘌呤二核苷酸).

糖酵解的定义。

微课：糖酵解

Overview of glycolysis.

Glycolysis is a definite sequence of ten reactions involving ten intermediate compounds(中间产物) (one of the steps involves two intermediates(中间体)). The intermediates provide entry points(切入点) to glycolysis. For example, most monosaccharides(单糖), such as fructose, glucose, and

糖酵解有十步反应,包含十个中间产物。中间体可以作为糖酵解的切入点,如二羟丙酮磷酸可以作为甘油的来源,和脂肪酸结合生成脂肪。

galactose(半乳糖), can be converted to one of these intermediates. The intermediates may also be directly useful. For example, the intermediate dihydroxyacetone phosphate(二羟丙酮磷酸)(DHAP) is a source of the glycerol(甘油) that combines with fatty acids(脂肪酸) to form fat. It occurs, with variations(变异), in nearly all organisms(有机体), both aerobic(需氧的) and anaerobic(厌氧的). The wide occurrence(发生) of glycolysis indicates that it is one of the most ancient known metabolic pathways.

> 糖酵解是存在于所有生物体内的一种重要的代谢途径。

The most common type of glycolysis is the Embden-Meyerhof-Parnas pathway (EMP pathway, EMP途径), which was first discovered by Gustav Embden, Otto Meyerhof and Jakub Karol Parnas. Glycolysis also refers to other pathways, such as the Entner-Doudoroff pathway(脱氧酮糖酸途径) and various heterofermentative(异型发酵的) and homofermentative(同型发酵的) pathways. However, the discussion here will be limited to the Embden-Meyerhof pathway.

> 糖酵解最普遍的形式是EMP途径。

2.1.1 Overview 概述

The overall reaction of glycolysis is:

D-[Glucose] + 2 [NAD]$^+$ + 2 [ATP] + 2 [P]$_i$ ⟶
2 [Pyruvate] + 2 [NADH] + 2 H$^+$ + 2 [ATP] + 2 H$_2$O

> 此方程中,原子的平衡由磷酸基团维持。

The use of symbols in this equation makes it appear unbalanced with respect to oxygen atoms, hydrogen(氢) atoms, and charges(电荷). Atom balance is maintained by the two phosphate (Pi) groups(磷酸基团):

(1) each exists in the form of a hydrogen phosphate anion (HPO_4^{2-})(磷酸氢负离子), dissociating(解离) to contribute $2H^+$ overall.

(2) each liberates(释放) an oxygen atom when it binds to an ADP (adenosine diphosphate) molecule, contributing 2O overall.

Charges are balanced by the difference between ADP and ATP. In the cellular environment(细胞环境), all three hydroxy groups(羟基) of ADP dissociate into $-O^-$ and H^+, giving ADP^{3-}, and this ion(离子) tends to exist in an ionic bond(离子键) with Mg^{2+}, giving $ADPMg^-$. ATP behaves identically(相等地) except that it has four hydroxy groups, giving $ATPMg^{2-}$. When these differences along with the true charges on the two phosphate groups are considered together, the net charges(净电荷) of -4 on each side are balanced.

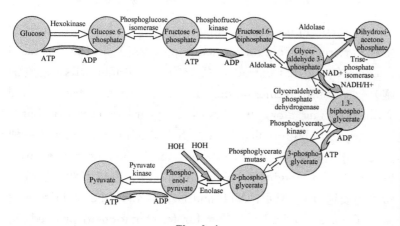

Glycolysis.

Hexokinase：己糖激酶
Phosphoglucose Isomerase：葡萄糖磷酸异构酶
Phosphofructokinase：磷酸果糖激酶
Aldolase：醛缩酶
Triosephosphate Isomerase：磷酸丙糖异构酶
Glyceraldehyde phosphate dehydrogenase：磷酸甘油醛脱氢酶
Phosphoglycerate kinase：磷酸甘油酸激酶
Phosphoglycerate mutase：磷酸甘油酸变位酶
Enolase：烯醇酶

For simple anaerobic fermentations(无氧发酵), the metabolism of one molecule of glucose to two molecules of pyruvate has a net yield(净生成) of two molecules of ATP. Most cells will then carry out further reactions to 'repay' the used NAD^+ and produce a final product of ethanol(乙醇) or lactic acid(乳酸). Many bacteria(细菌) use inorganic compounds(无机化合物) as hydrogen acceptors(受体) to regenerate(重新形成) the NAD^+. Cells performing aerobic respiration(有氧呼吸) synthesize much more ATP, but not as part of glycolysis. These further aerobic reactions use pyruvate and NADH+H$^+$ from glycolysis. Eukaryotic(真核的) aerobic respiration produces approximately 34 additional molecules of ATP for each glucose molecule, however most of these are produced by a vastly different mechanism(原理) to the substrate-level phosphorylation（底物水平磷酸化）in glycolysis.

真核有氧呼吸每分子葡萄糖额外产生34分子ATP,它们中大部分生成原理和糖酵解中的底物水平磷酸化不同。

The lower-energy production, per glucose, of anaerobic respiration relative to aerobic respiration, results in greater flux（流量）through the pathway under hypoxic (low-oxygen) conditions（低氧环境）, unless alternative sources of anaerobically-oxidizable substrates（厌氧性可氧化底物）, such as fatty acids, are found.

相对于有氧呼吸,无氧呼吸的不同之处：产生较少的能量,导致更大的通量。

2.1.2 Elucidation of the pathway 代谢途径说明

In 1860, Louis Pasteur discovered that microorganisms(微生物) are responsible for fermentation（发酵）. In 1897, Eduard Buchner found that extracts（提取物）of certain cells（某些细胞）can cause fermentation. In 1905, Arthur Harden and William Young along with Nick Sheppard determined that a heat-sensitive high-molecular-weight subcellular fraction（一种热敏高分子量亚细胞组分）(the enzymes) and a heat-insensitive low-molecular-weight cytoplasm fraction（一种热不敏感低分子量细胞质组分）(ADP, ATP and NAD$^+$ and other cofactors 辅酶因子) are required together for fermentation to proceed(发生). The details of the pathway were eventually determined by 1940, with a major input from Otto Meyerhof and some years later by Luis Leloir. The biggest difficulties in determining the intricacies(错综复杂的事物) of the pathway were due to the very short lifetime and low steady-state concentrations(不稳定的浓度) of the intermediates(中间产物) of the fast glycolytic（糖分解的）reactions. The first five steps are regarded as the preparatory (or investment(投入)) phase, since they consume energy to convert the glucose into two three-carbon sugar phosphates (G3P).

糖酵解途径的发现历程。

2.1.3 Sequence of reactions 反应顺序

(1) Preparatory phase 准备阶段

The first step in glycolysis is phosphorylation(磷酸化) of glucose by a family of enzymes called hexokinases(己糖激酶) to form glucose 6-phosphate (G6P)（葡萄糖-6-磷酸）. This reaction consumes ATP, but it acts to keep the glucose concentration low, promoting continuous transport of glucose into the cell through the plasma membrane transporters(质膜转运体). In addition, it blocks(阻碍) the glucose from leaking out - the cell lacks transporters for G6P, and free diffusion(自由扩散) out of the cell is prevented due to the charged nature (带电性质) of G6P. Glucose may alternatively be from the phosphorolysis(磷酸分解) or hydrolysis(水解) of intracellular (细胞内) starch(淀粉) or glycogen(糖原).

In animals, an isozyme(同工酶) of hexokinase called glucokinase(葡糖激酶) is also used in the liver(肝脏), which has a much lower affinity(亲和力) for glucose (K_m in the vicinity(邻近) of normal glycemia(血糖)), and differs in regulatory properties(调节性质). The different substrate affinity and alternate regulation(交替调节) of this enzyme are a reflection of the role of the liver in maintaining blood sugar levels.

Cofactors: Mg^{2+}

D-Glucose(Glc)	Hexokinase(HK) a transferase	α-D-Glucose-6-phosphate (G6P)

糖酵解第二步反应过程，即 G6P 在葡萄糖磷酸异构酶的作用下转变成 F6P。

G6P is then rearranged into fructose 6-phosphate (F6P) by glucose phosphate isomerase(葡萄糖磷酸异构酶). Fructose can also enter the glycolytic pathway（糖酵解途径） by phosphorylation(磷酸化) at this point. The change in structure is an isomerization(异构化), in which the G6P has been converted to F6P. The reaction requires an enzyme, phosphohexose isomerase(磷酸己糖异构酶), to proceed. This

此反应是可逆的。

reaction is freely reversible（可逆的）under normal cell conditions. However, it is often driven forward because of a low concentration of F6P, which is constantly consumed during the next step of glycolysis. Under conditions of high F6P concentration, this reaction readily runs in reverse. This phenomenon can be explained through Le Chatelier's Principle (勒夏特列原理). Isomerization to a keto sugar（酮糖） is necessary for carbanion stabilization(负碳离子稳定) in the fourth reaction step (below).

Phosphoglucose isomerase:磷酸葡萄糖异构酶

Fructose 6 - phosphate (F6P):果糖-6-磷酸

α-D-Glucose 6-phosphate (G6P)	Phosphoglucose isomerase An isomerase	β-D-Fructose 6-phosphat (F6P)

糖酵解第三步反应阶段受磷酸果糖激酶 1 催化，是不可逆的，且是一步限速反应。

The energy expenditure(消耗) of another ATP in this step is justified in 2 ways: The glycolytic process (up to this step) is now irreversible, and the energy supplied destabilizes the molecule. Because the reaction catalyzed（催化） by Phosphofructokinase 1(磷酸果糖激酶 1, PFK-1) is coupled to the hydrolysis(水解) of ATP, an energetically favorable step, it is, in essence, irreversible, and a different pathway must be used to do the reverse conversion(转换) during gluconeogenesis

（糖异生）. This makes the reaction a key regulatory point（see below, 调节点）. This is also the rate-limiting（限速）step.

Furthermore, the second phosphorylation event is necessary to allow the formation of two charged groups (rather than only one) in the subsequent step（后续步骤）of glycolysis, ensuring the prevention of free diffusion（自由扩散）of substrates out of the cell. The same reaction can also be catalyzed by pyrophosphate-dependent phosphofructokinase (PFP or PPi‐PFK), which is found in most plants, some bacteria, archae, and protists（原生生物）, but not in animals. This enzyme uses pyrophosphate（PPi）（焦磷酸）as a phosphate donor（供体）instead of ATP. It is a reversible reaction, increasing the flexibility（灵活性）of glycolytic metabolism. A rarer ADP-dependent PFK enzyme variant（多样的）has been identified in archaean species.

Cofactors: Mg^{2+}

第二步磷酸化由PFP(or PPi‐PFK)催化，不能在动物体内发生。这个酶可以利用焦磷酸作为磷酸供体来取代ATP，是可逆反应，可以提高糖代谢的灵活性。

| β-D-Fructose 6-phosphate (F6P) | Phosphofructokinase (PFK‐1) a transferase | β-D-Fructose 1,6-bisphosphate (F1,6BP) |

Phosphofructokinase (PFK‐1): 磷酸果糖激酶

Fructose 1,6-bisphosphat (F1,6BP): 果糖-1,6-二磷酸

Destabilizing the molecule in the previous reaction allows the hexose ring（己糖环）to be split by aldolase（醛缩酶）into two triose sugars（三糖）, dihydroxyacetone phosphate（二羟丙酮磷酸）, a ketone（酮）, and glyceraldehyde 3-phosphate（甘油醛-3-磷酸）, an aldehyde（醛）. There are two classes of aldolases: class I aldolases, present in animals and plants, and class II aldolases, present in fungi（真菌）and bacteria; the two classes use different mechanisms（机制）in cleaving the ketose ring. Electrons delocalized（移位电子）in the carbon-carbon bond

己糖环在醛缩酶的作用下分成两个三糖，二羟丙酮磷酸，酮，甘油醛-3-磷酸。

两类醛缩酶存在于不同的生物体中，它们使酮糖环断裂的机制不同。

cleavage(断裂) associate with the alcohol group(醇基). The resulting carbanion(负碳离子) is stabilized by the structure of the carbanion itself via resonance charge distribution(共振电荷分布) and by the presence of a charged ion prosthetic group(辅助基团).

Fructose bisphosphate aldolase (ALDO):果糖二磷酸醛缩酶

D-Glyceraldehyde 3-phosphate(GADP):甘油醛-3-磷酸

Dihydroxyacetone phosphate (DHAP):二羟丙酮磷酸

β-D-fructose 1, 6-bisphosphate (F1,6BP)	Fructose-bisphosphate aldolase (ALDO)	D-Glyceraldehyde 3-phosphate(GADP)	Dihydroxyacetone phosphate(DHAP)

磷酸丙酮异构酶完成 DHAP 和 GADP 之间的互换。

Triosephosphate isomerase rapidly interconverts(互换) dihydroxyacetone phosphate with glyceraldehyde 3-phosphate (GADP) that proceeds further into glycolysis. This is advantageous, as it directs dihydroxyacetone phosphate down the same pathway as glyceraldehyde 3-phosphate, simplifying regulation.

Triosephosphate isomerase (TPI):磷酸丙糖异构酶

糖酵解的第二部分是能量获得阶段,这一阶段的特点是 ATP 和 NADH 的净生成。

Dihydroxyacetone phosphate (DHAP)	Triosephosphate isomerase (TPI) an isomerase	D-Glyceraldehyde 3-phosphate (GADP)

(2) Pay-off phase 偿还阶段

The second half of glycolysis is known as the pay-off

phase, characterised by a net gain(净生成) of the energy-rich molecules ATP and NADH. Since glucose leads to two triose sugars in the preparatory phase, each reaction in the pay-off phase occurs twice per glucose molecule. This yields 2 NADH molecules and 4 ATP molecules, leading to a net gain of 2 NADH molecules and 2 ATP molecules from the glycolytic pathway per glucose.

The triose sugars are dehydrogenated(脱氢) and inorganic phosphate(无机磷酸盐) is added to them, forming 1,3-bisphosphoglycerate(1,3-二磷酸甘油酸). The hydrogen(氢) is used to reduce(还原) two molecules of NAD^+, a hydrogen carrier, to give $NADH + H^+$ for each triose. Hydrogen atom balance and charge balance are both maintained because the phosphate (P_i) group actually exists in the form of a hydrogen phosphate anion (HPO_4^{2-}), which dissociates to contribute the extra H^+ ion and gives a net charge of −3 on both sides.

三碳糖脱氢,与无机磷酸盐结合,形成1,3-BPG。

氢在此反应中用于还原2分子的NAD^+,磷酸基团以HPO_4^{2-}的形式维持反应中氢离子平衡和电荷平衡。

Glyceraldehyde 3-phosphate (GADP)	Glyceraldehyde phosphate dehydrogenase (GAPDH)	D-1, 3-Bisphosphoglycerate (1,3-BPG)

Glyceraldehyde phosphate Dehydrogenase (GAPDH):磷酸甘油醛脱氢酶

1,3-Bisphosphoglycerate (1,3-BPG):1,3-二磷酸甘油酸

This step is the enzymatic transfer(酶转移) of a phosphate group from 1,3-bisphosphoglycerate to ADP by phosphoglycerate kinase(磷酸甘油酸激酶), forming ATP and 3-phosphoglycerate (3-磷酸甘油酸). At this step, glycolysis has reached the break-even point(等值点): 2 molecules of ATP were consumed, and 2 new molecules have now been synthesized. This step, one of the two substrate-level phosphorylation steps

这步反应是1,3-BPG的磷酸基团在磷酸甘油酸激酶的作用下转移到ADP上,生成ATP和3-磷酸甘油酸的过程。

（底物水平磷酸化），requires ADP；thus，when the cell has plenty of ATP (and little ADP)，this reaction does not occur. Because ATP decays(衰减) relatively quickly when it is not metabolized，this is an important regulatory point(调节点) in the glycolytic pathway.

该步是糖酵解途径的调节点。

ADP actually exists as ADPMg$^-$, and ATP as ATPMg^{2-}, balancing the charges at -5 both sides.

Cofactors：Mg^{2+}

Phosphoglycerate kinase (PGK)：磷酸甘油酸激酶

3 - Phosphoglycerate(3 - PG)：3-磷酸甘油酸

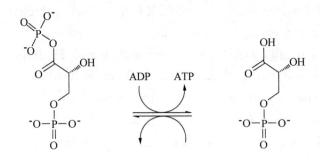

| 1,3 - Bisphosphoglycerate (1,3 - BPG) | Phosphoglycerate kinase (PGK) | 3 - Phosphoglycerate (3 - PG) |

3 - PG 在 PGM 作用下生成 2 - PG。

Phosphoglycerate mutase(磷酸甘油酸变位酶) now forms 2-phosphoglycerate from 3 - phosphoglycerate.

Phosphoglycerate mutase (PGM)：磷酸甘油酸变位酶

2 - Phosphoglycerate(2 - PG)：2-磷酸甘油酸

| 3 - Phosphoglycerate (3 - PG) | Phosphoglycerate mutase (PGM) | 2 - Phosphoglycerate (2 - PG) |

2 - PG 在烯醇酶的作用下生成磷酸烯醇丙酮酸。

Enolase(烯醇酶) next forms phosphoenolpyruvate(磷酸烯醇丙酮酸) from 2-phosphoglycerate.

Cofactors：2 Mg^{2+}：one "conformational(构象的)" ion to coordinate with(协调) the carboxylate group(羧基) of the substrate, and one "catalytic(催化的)" ion that participates in the dehydration(脱水).

| 2 - Phosphoglycerate (2 - PG) | Enolase (ENO) | Phosphoenolpyruvate (PEP) |

Enolase(ENO)：烯醇酶
Phosphoenolpyruvate (PEP)：磷酸烯醇丙酮酸

A final substrate-level phosphorylation now forms a molecule of pyruvate(丙酮酸) and a molecule of ATP by means of the enzyme pyruvate kinase(丙酮酸激酶). This serves as an additional regulatory step, similar to the phosphoglycerate kinase step.

最后一步反应在丙酮酸激酶的调解下通过底物水平磷酸化生成1分子的丙酮酸和1分子的ATP。

Cofactors：Mg^{2+}

| Phosphoenolpyruvate (PEP) | Pyruvate kinase (PK) | Pyruvate (Pyr) |

Pyruvate kinase (PK)：丙酮酸激酶
Pyruvate (Pyr)：丙酮酸

2.1.4 Regulation 调节

Glycolysis is regulated by slowing down or speeding up certain steps in the glycolysis pathway. This is accomplished by inhibiting or activating the enzymes(抑制或激活参与的酶) that are involved. The steps that are regulated may be determined by calculating(计算) the change in free energy(自由能), ΔG, for each step. If a step's products(产物) and reactants(反应物) are in equilibrium(平衡), then the step is assumed to not be

糖酵解的调节是通过减慢或加速糖酵解途径中的某些步骤来实现的。

regulated. Since the change in free energy is zero for a system at equilibrium, any step with a free energy change near zero is not being regulated. If a step is being regulated, then that step's enzyme is not converting reactants(反应物) into products as fast as it could, resulting in a build-up of reactants, which would be converted to products if the enzyme were operating faster. Since the reaction is thermodynamically(热力学的) favorable, the change in free energy for the step will be negative (负的). A step with a large negative change in free energy is assumed to be regulated.

自由能出现较大的负变化的步骤被认为是受到调节。

Concentrations of metabolites in erythrocytes.

Compound	Concentration/mM
glucose 葡萄糖	5.0
glucose-6-phosphate 葡萄糖-6-磷酸	0.083
fructose-6-phosphate 果糖-6-磷酸	0.014
fructose-1,6-bisphosphate 果糖-1,6-二磷酸	0.031
dihydroxyacetone phosphate 二羟丙酮磷酸	0.14
glyceraldehyde-3-phosphate 甘油醛-3-磷酸	0.019
1,3-bisphosphoglycerate 甘油酸-1,3-二磷酸	0.001
2,3-bisphosphoglycerate 甘油酸-2,3-二磷酸	4.0
3-phosphoglycerate 甘油酸-3-磷酸	0.12
2-phosphoglycerate 甘油酸-2-磷酸	0.03
phosphoenolpyruvate 烯醇丙酮酸磷酸	0.023
pyruvate 丙酮酸	0.051
ATP	1.85
ADP	0.14
P_i	1.0

自由能变化的计算方法。计算时需要知道代谢浓度。
NAD^+ 和 NADH 的比率大约是1,所以它们的浓度在反应系数中可以抵消。

The change in free energy, ΔG, for each step in the glycolysis pathway can be calculated using $\Delta G = \Delta G^{\circ\prime} + RT \ln Q$, where Q is the reaction quotient(系数). This requires knowing the concentrations of the metabolites. All of these values are available for erythrocytes(红细胞), with the exception of the concentrations of NAD^+ and NADH. The ratio

（比率）of NAD⁺ to NADH is approximately 1, which results in these concentrations canceling out（抵消）in the reaction quotient.（Since NAD⁺ and NADH occur on opposite sides of the reactions, one will be in the numerator(分子) and the other in the denominator(分母).

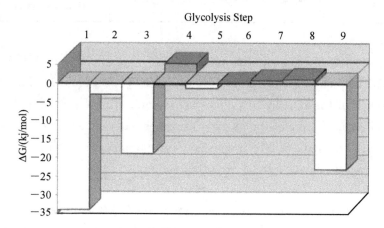

The change in free energy for each step of glycolysis estimated from the concentration of metabolites in an erythrocyte(红细胞).

Using the measured concentrations of each step, and the standard free energy changes, the actual free energy change can be calculated.

根据每一步反应的浓度和标准自由能变化,就可以计算出实际的自由能变化。

Change in free energy for each step of glycolysis.

Step	Reaction	$\Delta G^{\circ\prime}/$(kJ/mol)	$\Delta G/$(kJ/mol)
1	glucose + ATP^{4-} ⟶ glucose-6-phosphate^{2-} + ADP^{3-} + H$^+$	−16.7	−34
2	glucose-6-phosphate^{2-} ⟶ fructose-6-phosphate^{2-}	1.67	−2.9
3	fructose-6-phosphate^{2-} + ATP^{4-} ⟶ fructose-1,6-bisphosphate^{4-} + ADP^{3-} + H+	−14.2	−19
4	fructose-1,6-bisphosphate^{4-} ⟶ dihydroxyacetone phosphate^{2-} + glyceraldehyde-3-phosphate^{2-}	23.9	−0.23
5	dihydroxyacetone phosphate^{2-} ⟶ glyceraldehyde-3-phosphate^{2-}	7.56	2.4
6	glyceraldehyde-3-phosphate^{2-} + Pi^{2-} + NAD⁺ ⟶ 1,3-bisphosphoglycerate^{4-} + NADH + H⁺	6.30	−1.29
7	1,3-bisphosphoglycerate^{4-} + ADP^{3-} ⟶ 3-phosphoglycerate^{3-} + ATP^{4-}	−18.9	0.09

(续表)

Step	Reaction	$\Delta G^{\circ\prime}$/(kJ/mol)	ΔG/(kJ/mol)
8	3-phosphoglycerate^{3-} ⟶ 2-phosphoglycerate^{3-}	4.4	0.83
9	2-phosphoglycerate^{3-} ⟶ phosphoenolpyruvate^{3-} + H$_2$O	1.8	1.1
10	phosphoenolpyruvate^{3-} + ADP^{3-} + H$^+$ ⟶ pyruvate$^-$ + ATP^{4-}	−31.7	−23.0

糖酵解反应中,有七步反应是平衡的,有三步反应是不平衡的。

From measuring the physiological concentrations(生理浓度) of metabolites in a erythrocyte(红细胞) it seems that about seven of the steps in glycolysis are in equilibrium(平衡) for that cell type. Three of the steps — the ones with large negative free energy changes — are not in equilibrium and are referred to as *irreversible*; such steps are often subject to regulation.

反应5是副反应,可以减少或增加中间产物甘油醛-3-磷酸的浓度。在磷酸甘油醛异构酶催化下可以将中间产物转化成二羟丙酮磷酸,这步反应的速率很快,所以被认为是平衡的。但 ΔG 不是0暗示着红细胞中的实际浓度是不确定的。

Step 5 in the figure is shown behind the other steps, because that step is a side-reaction(副反应) that can decrease or increase the concentration of the intermediate (中间的) glyceraldehyde-3-phosphate. That compound is converted to dihydroxyacetone phosphate by the enzyme triose phosphate isomerase, which is a catalytically perfect enzyme; its rate is so fast that the reaction can be assumed to be in equilibrium. The fact that ΔG is not zero indicates that the actual concentrations in the erythrocyte are not accurately known.

(1) Biochemical logic 生化逻辑

多个调节点可以让中间体通过其他过程进入和离开糖酵解途径。
如第一步反应中,葡萄糖-6-磷酸可以转变成糖原或淀粉,通过可逆反应和降解又可以还原。这样产生的葡萄糖-6-磷酸能在第一个控制点后进入糖酵解。

The existence of more than one point of regulation indicates that intermediates between those points enter and leave the glycolysis pathway by other processes. For example, in the first regulated step, hexokinase converts glucose into glucose-6-phosphate. Instead of continuing through the glycolysis pathway, this intermediate can be converted into glucose storage molecules, such as glycogen(糖原) or starch(淀粉). The reverse reaction, breaking down, e. g., glycogen, produces mainly glucose-6-phosphate; very little free glucose is formed in the reaction. The glucose-6-phosphate so produced can enter glycolysis after the first control point.

In the second regulated step (the third step of glycolysis), phosphofructokinase converts fructose-6-phosphate into fructose-1,6-bisphosphate, which then is converted into glyceraldehyde-3-phosphate and dihydroxyacetone phosphate. The dihydroxyacetone phosphate can be removed from glycolysis by conversion into glycerol-3-phosphate, which can be used to form triglycerides (甘油三酯). On the converse, triglycerides can be broken down into fatty acids and glycerol; the latter, in turn, can be converted into dihydroxyacetone phosphate, which can enter glycolysis after the second control point.

(2) Three regulated enzymes 三个调节酶

The three regulated enzymes are hexokinase, phosphofructokinase, and pyruvate kinase.

The flux(流量) through the glycolytic pathway is adjusted in response to conditions both inside and outside the cell. The rate in liver is regulated to meet major cellular needs：(1) the production of ATP, (2) the provision(供应) of building blocks for biosynthetic reactions(生物合成反应), and (3) to lower blood glucose(降低血糖), one of the major functions of the liver. When blood sugar falls, glycolysis is halted(停止) in the liver to allow the reverse process, gluconeogenesis. In glycolysis, the reactions catalyzed by hexokinase, phosphofructokinase, and pyruvate kinase are effectively irreversible in most organisms. In metabolic pathways, such enzymes are potential sites of control, and all three enzymes serve this purpose in glycolysis.

Hexokinase 己糖激酶

In animals, regulation of blood glucose levels by the pancreas(胰腺) in conjunction with(协同) the liver is a vital part of homeostasis(体内平衡). In liver cells, extra G6P (glucose-6-phosphate) may be converted to G1P for conversion to glycogen, or it is alternatively converted by glycolysis to acetyl-CoA and then citrate(柠檬酸盐). Excess citrate is exported to the cytosol(胞液), where ATP citrate lyase(柠檬酸

式。G6P 不阻止催化葡萄糖被磷酸化为 G6P，因此己糖激酶活力低时，可以让葡萄糖转变成糖原、脂肪酸和胆固醇。处于低血糖症时，糖原又可以变回 G6P，再次变为葡萄糖，这个可逆反应对维持血糖水平至关重要，也是脑功能的关键，因为大脑可在多种环境下利用葡萄糖作为能源。

裂合酶) will regenerate acetyl-CoA and OAA. The acetyl-CoA is then used for fatty acid synthesis and cholesterol（胆固醇）synthesis, two important ways of utilizing（利用）excess glucose when its concentration is high in blood. Liver contains both hexokinase and glucokinase; the latter catalyses the phosphorylation of glucose to G6P and is not inhibited by G6P. Thus, it allows glucose to be converted into glycogen, fatty acids, and cholesterol even when hexokinase activity is low. This is important when blood glucose levels are high. During hypoglycemia（低血糖症）, the glycogen can be converted back to G6P and then converted to glucose by the liver-specific enzyme glucose 6-phosphatase. This reverse reaction is an important role of liver cells to maintain blood sugars levels during fasting. This is critical（关键的）for brain function, since the brain utilizes glucose as an energy source under most conditions.

Phosphofructokinase 磷酸果糖激酶

磷酸果糖激酶是糖酵解途径重要的控制点，因为它是不可逆反应中的一个，并且拥有关键的别构调节物：AMP 和 F2,6BP。

Phosphofructokinase is an important control point in the glycolytic pathway, since it is one of the irreversible steps and has key allosteric effectors（别构效应物）, AMP and fructose 2,6-bisphosphate（F2,6BP）.

Fructose 2,6-bisphosphate（F2,6BP）is a very potent activator（有效的催化剂）of phosphofructokinase（PFK-1）, which is synthesised when F6P is phosphorylated by a second phosphofructokinase（PFK2）. In liver, when blood sugar is low and glucagon（胰高血糖素）elevates cAMP, PFK2 is phosphorylated by protein kinase A. The phosphorylation inactivates（抑制剂）PFK2, and another domain（结构域）on this protein becomes active as fructose 2,6-bisphosphatase, which converts F2,6BP back to F6P. Both glucagon and epinephrine（肾上腺素）cause high levels of cAMP in the liver. The result of lower levels of liver fructose-2,6-bisphosphate is a decrease in activity of phosphofructokinase and an increase in activity of fructose 1,6-bisphosphatase, so that gluconeogenesis (in essence, "glycolysis in reverse") is favored. This is consistent with the role of the liver in such situations, since the

肝脏中，当血糖降低和胰高血糖素提升 cAMP，PFK2 就会被蛋白激酶 A 磷酸化。

胰高血糖素和肾上腺素都可以提升肝脏中 cAMP 的水平。肝脏中 F-2,6-BP 浓度下降，导致磷酸果糖激酶酶活下降和 F-1,6-二磷酸酶的酶活上升，从而有利于糖异生的途径。

response of the liver to these hormones is to release glucose to the blood.

ATP competes with AMP for the allosteric effector site on the PFK enzyme. ATP concentrations in cells are much higher than those of AMP, typically 100-fold higher, but the concentration of ATP does not change more than about 10% under physiological conditions(生理条件), whereas a 10% drop in ATP results in a 6-fold increase in AMP. Thus, the relevance (关联) of ATP as an allosteric effector is questionable. An increase in AMP is a consequence of a decrease in energy charge in the cell.

Citrate inhibits phosphofructokinase when tested in vitro (在体外) by enhancing the inhibitory effect of ATP. However, it is doubtful that this is a meaningful effect in vivo(在体内), because citrate in the cytosol is utilized mainly for conversion to acetyl-CoA for fatty acid and cholesterol synthesis.

Pyruvate kinase 丙酮酸激酶

This enzyme catalyzes the last step of glycolysis, in which pyruvate and ATP are formed. Regulation of this enzyme is discussed in the main topic, pyruvate kinase.

2.1.5　Post-glycolysis processes 糖酵解后的过程

The overall process of glycolysis is:
glucose$+2NAD^++2ADP+2P_i \longrightarrow 2$pyruvate$+2NADH+2H^++2ATP+2H_2O$

If glycolysis were to continue indefinitely(无限地), all of the NAD^+ would be used up, and glycolysis would stop. To allow glycolysis to continue, organisms must be able to oxidize (使氧化) NADH back to NAD^+.

(1) Fermentation 发酵

One method of doing this is to simply have the pyruvate do

the oxidation; in this process, the pyruvate is converted to lactate(乳酸盐) in a process called lactic acid fermentation:

$$pyruvate + NADH + H^+ \longrightarrow lactate + NAD^+$$

在细菌和动物体内产生发酵的原因和方式。

This process occurs in the bacteria involved in making yogurt(酸奶) (the lactic acid causes the milk to curdle(凝固)). This process also occurs in animals under hypoxic(含氧量低的) (or partially-anaerobic) conditions, found, for example, in overworked muscles(过度疲劳的肌肉) that are starved of oxygen, or in infarcted(梗塞的) heart muscle cells. In many tissues, this is a cellular last resort for energy; most animal tissue cannot maintain anaerobic respiration for an extended length of time(长时间内). Some organisms, such as yeast(酵母), convert NADH back to NAD^+ in a process called ethanol fermentation(酒精发酵). In this process, the pyruvate is converted first to acetaldehyde(乙醛) and carbon dioxide(二氧化碳), then to ethanol(乙醇).

酵母酒精发酵的方式。

厌氧发酵可以让单细胞生物利用糖酵解作为他们唯一的能量来源。

Lactic acid fermentation and ethanol fermentation can occur in the absence of oxygen. This anaerobic fermentation allows many single-cell organisms(单细胞生物) to use glycolysis as their only energy source.

(2) Anaerobic respiration 无氧呼吸

In the above two examples of fermentation, NADH is oxidized by transferring two electrons to pyruvate. However, anaerobic bacteria use a wide variety of compounds as the terminal electron acceptors in cellular respiration: nitrogenous compounds(含氮化合物), such as nitrates(硝酸盐) and nitrites(亚硝酸盐); sulfur compounds(含硫化合物), such as sulfates(硫酸盐), sulfites(亚硫酸盐), sulfur dioxide(二氧化硫), and elemental sulfur(硫元素); carbon dioxide; iron compounds(铁化合物); manganese compounds(锰化合物); cobalt compounds(钴化合物); and uranium compounds(铀化合物).

厌氧细菌无氧呼吸需要作为末端电子受体的化合物。

(3) Aerobic respiration 有氧呼吸

In aerobic organisms, a complex mechanism(机制) has been created to use the oxygen in air as the final electron

有氧呼吸的机制。

acceptor of respiration.

- First, pyruvate is converted to acetyl-CoA and CO_2 within the mitochondria(线粒体) in a process called pyruvate decarboxylation.

丙酮酸的脱羧反应。

- Second, the acetyl-CoA enters the citric acid cycle, also known as Krebs Cycle(三羧酸循环), where it is fully oxidized to carbon dioxide and water, producing yet more NADH.

乙酰辅酶A进入三羧酸循环彻底氧化。

- Third, the NADH is oxidized to NAD^+ by the electron transport chain(电子传递链), using oxygen as the final electron acceptor. This process creates a "hydrogen ion gradient(氢离子梯度)" across the inner membrane of the mitochondria.

NADH在电子传递链中传递,最终被氧化为NAD^+。

- Fourth, the proton gradient(质子梯度) is used to produce a large amount of ATP in a process called oxidative phosphorylation(氧化磷酸化).

质子梯度被用来产生大量ATP,叫作氧化磷酸化。

(4) Intermediates for other pathways 为其他途径提供中间产物

This article concentrates on the catabolic role of glycolysis with regard to converting potential chemical energy to usable chemical energy during the oxidation of glucose to pyruvate. However, many of the metabolites in the glycolytic pathway are also used by anabolic pathways(同化途径), and, as a consequence, flux through the pathway is critical to maintain a supply of carbon skeletons(碳架) for biosynthesis.

糖酵解在将葡萄糖氧化成丙酮酸的同时可以将潜在的化学能量转变成有用的化学能量。糖酵解途径中的代谢物还可用于同化途径。

In addition, not all carbon entering the pathway leaves as pyruvate and may be extracted(获取) at earlier stages to provide carbon compounds for other pathways.

有的碳还可以在前几个步骤中获得,为其他途径提供含碳化合物。

These metabolic pathways are all strongly reliant on glycolysis as a source of metabolites:

- Gluconeogenesis(糖异生)
- Lipid metabolism
- Pentose phosphate pathway(戊糖磷酸途径)
  Citric acid cycle, which in turn leads to:
- Amino acid synthesis(氨基酸合成)
- Nucleotide synthesis(核苷酸合成)

某些代谢途径依赖糖酵解的代谢物。

● Tetrapyrrole synthesis（四吡咯合成）

From an anabolic metabolism perspective（从合成代谢的角度来看），the NADH has a role to drive synthetic reactions, doing so by directly or indirectly reducing the pool of $NADP^+$ in the cell to NADPH, which is another important reducing agent for biosynthetic pathways in a cell.

2.1.6　Glycolysis in disease 疾病中的糖酵解

Genetic diseases 遗传病

Glycolytic mutations（突变）are generally rare due to importance of the metabolic pathway, this means that the majority of occurring mutations result in an inability for the cell to respire, and therefore cause the death of the cell at an early stage. However, some mutations are seen with one notable（显著的）example being Pyruvate kinase deficiency（丙酮酸激酶缺乏症）, leading to chronic hemolytic anemia（溶血性贫血）.

Cancer 癌症

Malignant rapidly-growing tumor cells（恶性肿瘤细胞）typically have glycolytic rates that are up to 200 times higher than those of their normal tissues of origin. This phenomenon（现象）was first described in 1930 by Otto Warburg（奥托·瓦博格）and is referred to as the Warburg effect. The Warburg hypothesis claims that cancer is primarily caused by dysfunctionality（功能紊乱）in mitochondrial metabolism, rather than because of uncontrolled growth of cells. A number of theories have been advanced to explain the Warburg effect（瓦博格效应）.

This high glycolysis rate has important medical applications, as high aerobic glycolysis（有氧糖酵解）by malignant tumors is utilized clinically to diagnose（临床诊断）and monitor treatment responses of cancers by imaging uptake of $2-^{18}F-2-$ deoxyglucose (FDG, a radioactive modified hexokinase substrate（放射性修饰己糖激酶底物）with positron emission tomography（PET，正电子发射断层成像）.

2.2 Gluconeogenesis 糖异生

Gluconeogenesis（糖异生）is a metabolic pathway that results in the generation of glucose from non-carbohydrate carbon substrates（五碳底物）such as lactate, glycerol（甘油）, and glucogenic amino acids（生糖氨基酸）.

糖异生的定义。

It is one of the two main mechanisms（机制）humans and many other animals use to keep blood glucose levels from dropping too low (hypoglycemia，低血糖症). The other means of maintaining blood glucose levels is through the degradation of glycogen (glycogenolysis). Gluconeogenesis is a ubiquitous（普遍存在的）process, present in plants, animals, fungi, bacteria, and other microorganisms. In animals, gluconeogenesis takes place mainly in the liver and, to a lesser extent, in the cortex（皮质）of kidneys. This process occurs during periods of fasting（禁食）, starvation, low-carbohydrate diets, or intense exercise and is highly endergonic. For example, the pathway leading from phosphoenolpyruvate（磷酸烯醇丙酮酸）to glucose-6-phosphate requires 4 molecules of ATP and 2 molecules of GTP. Gluconeogenesis is often associated with ketosis. Gluconeogenesis is also a target of therapy（治疗）for type II diabetes, such as metformin（二甲双胍）, which inhibits glucose formation and stimulates（刺激）glucose uptake by cells.

糖异生是动物和人维持血糖水平的两种主要机制之一。

糖异生在动植物和微生物中普遍存在。

糖异生在动物内发生的几种情形和例子。

糖异生可用于治疗糖尿病Ⅱ。

2.2.1 Entering the pathway 进入途径

Lactate is transported back to the liver where it is converted into pyruvate by the Cori cycle using the enzyme lactate dehydrogenase（乳酸脱氢酶）. Pyruvate, the first designated substrate of the gluconeogenic pathway, can then be used to generate glucose.

乳酸被传送回肝脏转变成丙酮酸，丙酮酸又经糖异生形成葡萄糖。

All citric acid cycle intermediates, through conversion to

柠檬酸循环中间产物可通过转变为草酰乙酸，除了赖氨酸和亮氨酸以外的氨基酸以及甘油，从而作为糖异生的底物。氨基酸通过转氨作用或去氨基作用可以使碳骨架直接或间接得进入三羧酸循环。

oxaloacetate(草酰乙酸), amino acids other than lysine(赖氨酸) or leucine(亮氨酸), and glycerol can also function as substrates for gluconeogenesis. Transamination(转氨作用) or deamination (去氨基) of amino acids facilitates entering of their carbon skeleton into the cycle directly (as pyruvate or oxaloacetate), or indirectly via the citric acid cycle.

脂肪酸在动物体内是否能够转变成葡萄糖是生物化学中一个长期质疑的问题。

Whether fatty acids can be converted into glucose in animals has been a longstanding question in biochemistry. It is known that odd-chain fatty acids can be oxidized to yield propionyl CoA(丙酰辅酶 A), a precursor for succinyl CoA(琥珀酰辅酶 A), which can be converted to pyruvate and enter into gluconeogenesis. In plants, to be specific, in seedlings(幼苗), the glyoxylate cycle(乙醛酸循环) can be used to convert fatty acids (acetate) into the primary carbon source of the organism. The glyoxylate cycle produces four-carbon dicarboxylic acids(二羧酸) that can enter gluconeogenesis.

在植物体中脂肪酸的去向。

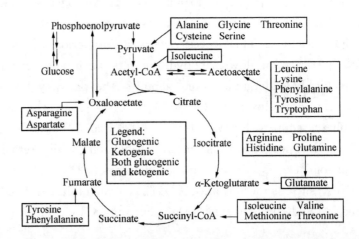

蛋白质氨基酸的分解代谢。

其中生糖氨基酸可以进入糖异生途径。

Catabolism of proteinogenic amino acids. Amino acids are classified according to the abilities of their products to enter gluconeogenesis: Glucogenic amino acids have this ability. Ketogenic amino acids do not. These products may still be used for ketogenesis or lipid synthesis. Some amino acids are catabolized into both glucogenic and ketogenic products.

很多哺乳动物体内发现苹果酸合酶的基因密码，但胎盘哺乳动物没有。异柠檬酸裂合酶的基因只在线虫里发现。

In 1995, researchers identified the glyoxylate cycle in nematodes(线虫). In addition, the glyoxylate enzymes malate synthase(苹果酸合酶) and isocitrate lyase(异柠檬酸裂合酶) have been found in animal tissues. Genes coding for malate synthase gene have been identified in other metazoans(多细胞动物) including arthropods(节肢动物), echinoderms(棘皮动物), and even some vertebrates(脊椎动物). Mammals(哺乳动物) found to possess these genes include monotremes(单孔目动物)(platypus, 鸭嘴兽) and marsupials(有袋类)(opossum, 负鼠) but not placental mammals(胎盘哺乳动物). Genes for isocitrate lyase are found only in nematodes, in which, it is apparent, they originated in horizontal gene transfer(基因水平转移) from bacteria.

The existence of glyoxylate cycles in humans has not been established, and it is widely held that fatty acids cannot be converted to glucose in humans directly. However, carbon-14 has been shown to end up in glucose when it is supplied in fatty acids. Despite these findings, it is considered unlikely that the 2-carbon acetyl-CoA derived from(来源于) the oxidation of fatty acids would produce a net yield of glucose via the citric acid cycle. However, it is possible that, with additional sources of carbon via other pathways, glucose could be synthesized from acetyl-CoA. In fact, it is known that Ketone bodies(酮体), β-hydroxybutyrate(羟丁酸) in particular, can be converted to glucose at least in small amounts (β - hydroxybutyrate to acetoacetate(乙酰乙酸) to acetone(丙酮) to propanediol(丙二醇) to pyruvate to glucose).

Glycerol, which is a part of the triacylglycerol molecule, can be used in gluconeogenesis.

2.2.2 Location 定位

In humans, gluconeogenesis is restricted to(局限于) the liver and to a lesser extent the kidney.

In all species, the formation of oxaloacetate(草酰乙酸) from pyruvate and TCA cycle intermediates is restricted to the mitochondrion, and the enzymes that convert PEP to glucose are found in the cytosol. The location of the enzyme that links these two parts of gluconeogenesis by converting oxaloacetate to

运输 PEP 的线粒体膜是通过专用的运输蛋白；然而乙醛酸循环不存在这样的蛋白质。

PEP，PEP carboxykinase（丙酮酸羧基酶），is variable by species: it can be found entirely within the mitochondria, entirely within the cytosol, or dispersed evenly between the two, as it is in humans. Transport of PEP across the mitochondrial membrane is accomplished by dedicated transport proteins; however, no such proteins exist for oxaloacetate.

当缺乏线粒体内的 PEP 时，为了使糖异生继续，草酰乙酸必须转变为苹果酸和天冬氨酸。

Therefore species that lack intra-mitochondrial PEP, oxaloacetate must be converted into malate or asparate（天冬氨酸）, exported from the mitochondrion, and converted back into oxaloacetate in order to allow gluconeogenesis to continue.

2.2.3 Pathway 途径

糖异生途径由 11 个酶催化反应组成，发生在线粒体或细胞质，依赖于底物。许多反应都是糖酵解的可逆步骤。

Gluconeogenesis is a pathway consisting of eleven enzyme-catalyzed reactions. The pathway can begin in the mitochondria（线粒体）or cytoplasm（细胞质）, depending on the substrate being used. Many of the reactions are the reversible steps found in glycolysis.

糖异生始于线粒体中丙酮酸羧化为草酰乙酸。

● Gluconeogenesis begins in the mitochondria with the formation of oxaloacetate through carboxylation（羧化作用）of pyruvate. This reaction also requires one molecule of ATP, and is catalyzed by pyruvate carboxylase（羧化酶）. This enzyme is stimulated by high levels of acetyl-CoA (produced in β-oxidation in the liver) and inhibited by high levels of ADP.

草酰乙酸的还原。

● Oxaloacetate is reduced to malate using NADH, a step required for transport out of the mitochondria.

苹果酸的氧化。

● Malate is oxidized to oxaloacetate using NAD^+ in the cytoplasm, where the remaining steps of gluconeogenesis occur.

草酰乙酸的脱羧和磷酸化。

● Oxaloacetate is decarboxylated and phosphorylated to produce phosphoenolpyruvate by phosphoenolpyruvate carboxykinase. One molecule of GTP is hydrolyzed（水解）to GDP during this reaction.

接着是糖酵解的逆步骤，然而 F-1,6-BP 转变成 F6P，是果糖-1,6-二糖磷酸酶催化的，这是糖异生中的一步限速反应。

● The next steps in the reaction are the same as reversed glycolysis. However, fructose-1,6-bisphosphatase converts fructose-1,6-bisphosphate to fructose 6-phosphate, requiring one water molecule and releasing one phosphate. This is also the rate-limiting step of gluconeogenesis.

- Glucose-6-phosphate is formed from fructose 6-phosphate by phosphoglucoisomerase(磷酸葡糖异构酶). Glucose-6-phosphate can be used in other metabolic pathways or dephosphorylated to free glucose. Whereas free glucose can easily diffuse in and out of the cell, the phosphorylated form (glucose-6-phosphate) is locked in the cell, a mechanism by which intracellular (胞内的) glucose levels are controlled by cells.

- The final reaction of gluconeogenesis, the formation of glucose, occurs in the lumen(内腔) of the endoplasmic reticulum(内质网), where glucose-6-phosphate is hydrolyzed by glucose-6-phosphatase to produce glucose. Glucose is shuttled into(穿梭进入) the cytosol by glucose transporters located in the membrane of the endoplasmic reticulum.

2.2.4 Regulation 调节

While most steps in gluconeogenesis are the reverse of those found in glycolysis, three regulated and strongly exergonic reactions(放能反应) are replaced with more kinetically(动力学) favorable reactions. Hexokinase/glucokinase, phosphofructokinase, and pyruvate kinase enzymes of glycolysis are replaced with glucose-6-phosphatase, fructose-1,6-bisphosphatase, and PEP carboxykinase. This system of reciprocal(相互的) control allow glycolysis and gluconeogenesis to inhibit each other and prevent the formation of a futile cycle(无效循环).

The majority of the enzymes responsible for gluconeogenesis are found in the cytoplasm; the exceptions are mitochondrial pyruvate carboxylase and, in animals, phosphoenolpyruvate carboxykinase(磷酸烯醇丙酮酸羧激酶).

The latter exists as an isozyme(同工酶) located in both the mitochondrion and the cytosol. The rate of gluconeogenesis is ultimately controlled by the action of a key enzyme, fructose-1,6-bisphosphatase, which is also regulated through signal tranduction(转导) by cAMP and its phosphorylation.

大部分能影响糖异生的因素,都是通过抑制关键酶的活力或表达,乙酰辅酶A和柠檬酸可以激活糖异生的酶类。	Most factors that regulate the activity of the gluconeogenesis pathway do so by inhibiting the activity or expression of key enzymes. However, both acetyl CoA and citrate activate gluconeogenesis enzymes(柠檬酸激活糖异生的酶)(pyruvate carboxylase and fructose-1,6-bisphosphatase, respectively). Due to the reciprocal control of the cycle, acetyl-CoA and citrate also have inhibitory roles in the activity of pyruvate kinase. Global control of gluconeogenesis is mediated by
胰高血糖素对糖异生的调节。	glucagon(胰高血糖素)(released when blood glucose is low); it triggers phosphorylation of enzymes and regulatory proteins by Protein Kinase A (a cyclic AMP regulated kinase) resulting in inhibition of glycolysis and stimulation of gluconeogenesis, thus bringing blood glucose levels up.

2.3 Glycogen 糖原

糖原的定义。	Glycogen is a molecule that serves as the secondary long-term energy storage in animal and fungal cells(真菌细胞), with the primary energy stores being held in adipose tissue(脂肪组织). Glycogen is made primarily by the liver and the muscles, but can also be made by glycogenesis within the brain and stomach.
糖原结构和淀粉类似。	Glycogen is the analogue(类似物) of starch, a less branched glucose polymer(聚合物) in plants, and is sometimes referred to as animal starch, having a similar structure to amylopectin(支链淀粉). Glycogen is found in the form of granules(颗粒) in the cytosol/cytoplasm in many cell types, and plays an important role in the glucose cycle. Glycogen forms an
糖原的功能。	energy reserve(能源储备) that can be quickly mobilized(动员) to meet a sudden need for glucose, but one that is less compact than the energy reserves of triglycerides (lipids).
糖原在体内的分布和不同的作用。	In the liver hepatocytes(肝细胞), glycogen can compose up to eight percent of the fresh weight (100~120 g in an adult) soon after a meal. Only the glycogen stored in the liver can be

made accessible to other organs. In the muscles, glycogen is found in a low concentration (one to two percent of the muscle mass). However, the amount of glycogen stored in the body—especially within the muscles, liver, and red blood cells—mostly depends on physical training, basal metabolic rate, and eating habits(饮食习惯) such as intermittent fasting(间歇性禁食). Small amounts of glycogen are found in the kidneys, and even smaller amounts in certain glial cells(胶质细胞) in the brain and white blood cells. The uterus(子宫) also stores glycogen during pregnancy(怀孕) to nourish the embryo(孕育胚胎).

2.3.1 Function and regulation of liver glycogen 肝糖原的功能与调节

As a meal containing carbohydrates is eaten and digested (消化), blood glucose levels rise, and the pancreas secretes insulin(胰腺分泌胰岛素). Glucose from the portal vein(门静脉) enters liver cells (hepatocytes). Insulin acts on the hepatocytes to stimulate the action of several enzymes, including glycogen synthase. Glucose molecules are added to the chains of glycogen as long as both insulin and glucose remain plentiful. In this postprandial(餐后的) or "fed" state, the liver takes in more glucose from the blood than it releases.

After a meal has been digested and glucose levels begin to fall, insulin secretion is reduced, and glycogen synthesis stops. When it is needed for energy, glycogen is broken down and converted again to glucose. Glycogen phosphorylase is the primary enzyme of glycogen breakdown. For the next 8~12 hours, glucose derived from liver glycogen will be the primary source of blood glucose to be used by the rest of the body for fuel.

Glucagon(胰高血糖素) is another hormone produced by the pancreas, which in many respects serves as a counter-signal (数据信号) to insulin. In response to insulin level below normal (when blood levels of glucose begin to fall below the normal range), glucagon is secreted in increasing amounts to stimulate glycogenolysis and gluconeogenesis pathways.

2.3.2 Function and regulation of glycogen in muscle and other cells 肌肉和其他细胞中糖原的功能与调节

Muscle cell glycogen appears to function as an immediate reserve source of available glucose for muscle cells. Other cells that contain small amounts use it locally as well. Muscle cells lack the enzyme glucose-6-phosphatase, which is required to pass glucose into the blood, so the glycogen they store is destined for internal use(内部使用) and is not shared with other cells. (This is in contrast to liver cells, which, on demand, readily do break down their stored glycogen into glucose and send it through the blood stream(血流) as fuel for the brain or muscles). Glycogen is also a suitable storage substance(储能物质) due to its insolubility in water.

2.3.3 Glycogen depletion and endurance exercise 糖原耗竭和耐力锻炼

Long-distance athletes such as marathon runners, cross-country skiers(越野滑雪), and cyclists often experience glycogen depletion, where almost all of the athlete's glycogen stores are depleted after long periods of exertion(运用) without enough energy consumption. This phenomenon is referred to as "hitting the wall(撞墙效应)". In marathon runners, it normally happens around the 20-mile (32 km) point of a marathon, depending on the size of the runner and the race course.

Glycogen depletion can be forestalled(阻止) in four possible ways. First, during exercise carbohydrates with the highest possible rate of conversion(转换) to blood glucose per time (high GI) are ingested(摄取) continuously. The best possible outcome of this strategy replaces about 35% of glucose consumed at heart rates above about 80% of maximum. Second, through training, the body can be conditioned to burn fat earlier, faster, and more efficiently, sparing(节约)

carbohydrate use from all sources. Third, by consuming foods low on the glycemic index(血糖指数) for 12~18 hours before the event, the liver and muscles will store the resulting slow but steady stream of glucose as glycogen, instead of fat. This is process is known as carbohydrate loading. Finally, taking breaks while continuing to ingest carbohydrate with a high GI allows the body to catch-up with consumption, in part, by the continued conversion of fat into glucose in the liver.

When experiencing glycogen debt, athletes often experience extreme fatigue to the point that it is difficult to move. As a reference, the very best professional cyclists in the world will usually finish a 4~5 hr stage race right at the limit of glycogen depletion using the first 3 strategies. A study published in the Journal of Applied Physiology (online May 8, 2008) suggests that, when athletes ingest both carbohydrate and caffeine following exhaustive exercise, their glycogen is replenished more rapidly.

2.3.4 Disorders of glycogen metabolism 糖原代谢紊乱

The most common disease in which glycogen metabolism becomes abnormal is diabetes, in which, because of abnormal amounts of insulin, liver glycogen can be abnormally accumulated or depleted. Restoration of normal glucose metabolism usually normalizes glycogen metabolism as well.

In hypoglycemia caused by excessive(过多的) insulin, liver glycogen levels are high, but the high insulin level prevents the glycogenolysis necessary to maintain normal blood sugar levels. Glucagon is a common treatment for this type of hypoglycemia.

Various inborn errors of metabolism are caused by deficiencies of enzymes necessary for glycogen synthesis or breakdown. These are collectively referred to as glycogen storage diseases.

2.3.5 Synthesis 合成

Glycogen synthesis is, unlike its breakdown, endergonic (吸热的). This means that glycogen synthesis requires the input of energy. Energy for glycogen synthesis comes from UTP, which reacts with glucose-1-phosphate, forming UDP-glucose, in reaction catalysed by UDP-glucose pyrophosphorylase(尿苷二磷酸葡萄糖焦磷酸化酶). Glycogen is synthesized from monomers of UDP-glucose by the enzyme glycogen synthase(合酶), which progressively lengthens (逐步延长) the glycogen chain with (α1→4) bonded glucose.

As glycogen synthase can lengthen only an existing chain, the protein glycogenin is needed to initiate the synthesis of glycogen. The Glycogen-branching enzyme, amylo(淀粉) (α1→4) to (α1→6) transglycosylase (转糖苷酶), catalyzes the transfer of a terminal fragment of 6~7 glucose residues from a nonreducing end to the C-6 hydroxyl group (羟基) of a glucose residue deeper into the interior (内部) of the glycogen molecule. The branching enzyme can act upon only a branch having at least 11 residues, and the enzyme may transfer to the same glucose chain or adjacent(邻近的) glucose chains.

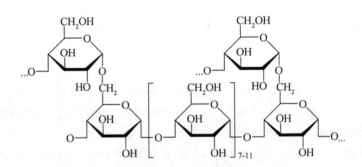

Glycogen Structure Segment.

2.3.6 Breakdown 分解

Glycogen is cleaved from the nonreducing ends of the chain by the enzyme glycogen phosphorylase to produce monomers(单

体) of glucose-1-phosphate, which is then converted to glucose 6-phosphate by phosphoglucomutase(葡萄糖磷酸变位酶). A special debranching enzyme(脱支酶) is needed to remove the alpha(1~6) branches in branched glycogen and reshape the chain into linear polymer(线型高分子). The G6P monomers produced have three possible fates:

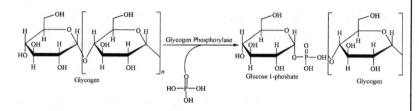

Action of Glycogen Phosphorylase on Glycogen.

- G6P can continue on the glycolysis pathway and be used as fuel.
- G6P can enter the pentose phosphate pathway(戊糖磷酸途径) via the enzyme Glucose-6-phosphate dehydrogenase to produce NADPH and 5-carbon sugars.
- In the liver and kidney, G6P can be dephosphorylated(去磷酸化) back to Glucose by the enzyme glucose 6-phosphatase. This is the final step in the gluconeogenesis pathway.

2.4 Pentose phosphate pathway 戊糖磷酸途径

The pentose phosphate pathway (also called the phosphogluconate pathway(磷酸葡萄糖途径) and the hexose monophosphate shunt(磷酸己糖支路)) is a process that generates NADPH and pentoses (5-carbon sugars). There are two distinct phases in the pathway. The first is the oxidative phase, in which NADPH is generated, and the second is the non-oxidative synthesis of 5-carbon sugars. This pathway is an alternative to glycolysis. While it does involve oxidation of glucose, its primary role is anabolic rather than catabolic. For most organisms, it takes place in the cytosol; in plants, most steps take place in plastids(叶绿体).

2.4.1 Phases 阶段

Oxidative phase 氧化阶段

In this phase, two molecules of NADP$^+$ are reduced to NADPH, utilizing the energy from the conversion of glucose-6-phosphate into ribulose 5-phosphate.

氧化阶段发生的反应。

Oxidative phase of pentose phosphate pathway. Glucose－6－phosphate（1）, 6－phosphoglucono-δ－lactone（2）, 6－phosphogluconate（3）, ribulose 5－phosphate（4）.

The entire set of reactions can be summarized as follows:

Reactants	Products	Enzyme	Description
Glucose 6－phosphate＋NADP$^+$	6－phosphoglucono-d－lactone＋NADPH	glucose 6－phosphate dehydrogenase	Dehydrogenation. The hemiacetal hydroxyl group（半缩醛羟基）located on carbon 1 of glucose 6－phosphate is converted into a carbonyl group, generating a lactone, and, in the process, NADPH is generated.
6－phosphoglucono-d－lactone＋HO$_2$	6－phosphogluconate ＋ H$^+$	6－phosphogluconolactonase	Hydrolysis（水解）
6－phosphogluconate ＋NADP$^+$	ribulose 5－phosphate＋NADPH＋CO$_2$	6－phosphogluconate dehydrogenase	Oxidative decarboxylation. NADP$^+$ is the electron acceptor, generating another moleculeof NADPH, a CO$_2$, and ribulose 5－phosphate.

The overall reaction for this process is:

Glucose 6-phosphate + 2 NADP$^+$ + H$_2$O ⟶ ribulose 5-phosphate + 2 NADPH + 2H$^+$ + CO$_2$

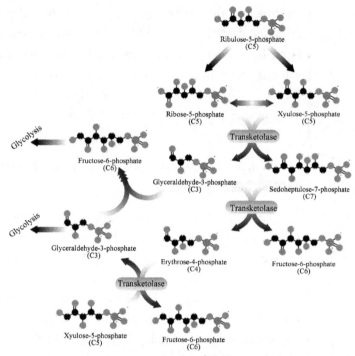

The pentose phosphate pathway's nonoxidative phase.

Non-oxidative phase 非氧化阶段

Reactions in non-oxidative phase.

Reactants	Products	Enzymes
ribulose 5-phosphate	→ ribose 5-phosphate	ribulose 5-phosphate isomerase
ribulose 5-phosphate	→ xylulose 5-phosphate	ribulose 5-phosphate 3-epimerase
xylulose 5-phosphate + ribose 5-phosphate	→ glyceraldehyde 3-phosphate + sedoheptulose 7-phosphate	transketolase
sedoheptulose 7-phosphate + glyceraldehyde 3-phosphate	→ erythrose 4-phosphate + fructose 6-phosphate	transaldolase
xylulose 5-phosphate + erythrose 4-phosphate	→ glyceraldehyde 3-phosphate + fructose 6-phosphate	transketolase

Net reaction:

3 ribulose-5-phosphate → 1 ribose-5-phosphate + 2 xylulose-5-phosphate → 2 fructose-6-phosphate + glyceraldehyde-3-phosphate

2.4.2 Regulation 调节

Glucose-6-phosphate dehydrogenase is the rate-controlling enzyme of this pathway. It is allosterically（别构调节）stimulated（诱导）by $NADP^+$. The ratio of $NADPH:NADP^+$ is normally about 100:1 in liver cytosol. This makes the cytosol a highly-reducing environment. An NADPH-utilizing pathway forms $NADP^+$, which stimulates Glucose-6-phosphate dehydrogenase to produce more NADPH.

2.4.3 Erythrocytes and the pentose phosphate pathway 红细胞和戊糖磷酸途径

Carbohydrates are metabolized in red blood cells mainly by glycolysis, the pentose phosphate pathway (PPP), and 2,3-bisphosphoglycerate（二磷酸甘油酸, 2,3-BPG) metabolism (refer to discussion of hemoglobin for the role of 2,3-BPG). Glycolysis provides ATP for membrane ion pumps and NADH for reduction of methemoglobin（高铁血红蛋白）. The PPP supplies the red blood cell with NADPH, which in turn maintains the reduced state of glutathione. The inability to maintain reduced glutathione in red blood cells leads to increased accumulation of peroxides（积累的过氧化物）, predominantly H_2O_2, that in turn results in a weakening of the cell membrane and concomitant hemolysis（伴随的溶血）. Accumulation of H_2O_2 also leads to increased rates of oxidation of hemoglobin（血红蛋白）to methemoglobin that also weakens the cell wall. Glutathione（谷胱甘肽）removes peroxides via the action of glutathione peroxidase（过氧化酶）. The PPP in erythrocytes is, in essence, the only pathway for these cells to produce NADPH. Any defect in the production of NADPH could, therefore, have profound effects on erythrocyte survival.

Several deficiencies（缺陷）in the level of activity (not function) of glucose-6-phosphate dehydrogenase have been observed to be associated with resistance to the malarial parasite（寄生虫）Plasmodium falciparum（恶性疟原虫）among individuals of Mediterranean（地中海人）and African descent（非洲后裔）. The basis for this resistance may be a weakening of the red cell membrane (the erythrocyte is the host cell for the parasite) such that it cannot sustain the parasitic(寄生的) life cycle long enough for productive growth.

2.5 Citric acid cycle 三羧酸循环

The citric acid cycle also known as the tricarboxylic acid cycle (TCA cycle), the Krebs cycle, or recently in certain former Soviet Bloc countries(苏联国家) the Szent-Györgyi-Krebs cycle is a series of enzyme-catalysed chemical reactions, which is of central importance in all living cells, especially those that use oxygen as part of cellular respiration. In eukaryotic cells, the citric acid cycle occurs in the matrix(基质) of the mitochondrion(线粒体). The components and reactions of the citric acid cycle were established by discovery of Vitamin C by Hungarian(匈牙利人) Nobel laureate Albert Szent-Györgyi and continued on to its complex metabolism into energy and metabolites by Nobel laureate(诺贝尔获奖者) Hans Adolf Krebs, a German born, Jewish refugee to Britain.

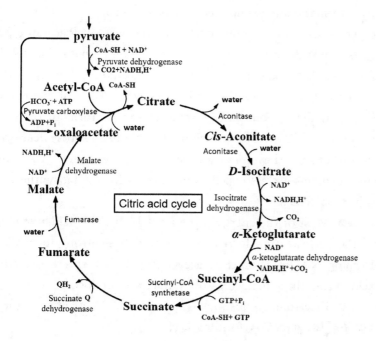

Overview of the citric acid cycle.

对于需氧生物，三羧酸循环是代谢途径的一部分，涉及碳水化合物、脂肪和蛋白质转变为二氧化碳和水，并产生可以利用的能量。

In aerobic organisms(需氧微生物), the citric acid cycle is part of a metabolic pathway involved in the chemical conversion of carbohydrates, fats and proteins into carbon dioxide and water to generate a form of usable energy. Other relevant reactions in the pathway include those in glycolysis and pyruvate(丙酮酸) oxidation before the citric acid cycle, and oxidative phosphorylation after it. In addition, it provides precursors(前体物) for many compounds including some amino acids and is therefore functional even in cells performing fermentation. Its centrality to many paths of biosynthesis suggest that it was one of the earliest formed parts of the cellular metabolic processes, and may have formed abiogenically(非生物成因).

2.5.1　A simplified view of the process 过程简介

三羧酸循环的起始阶段。

● The citric acid cycle begins with the transfer of a two-carbon acetyl group(乙酰基团) from acetyl-CoA to the four-carbon acceptor compound (oxaloacetate,草酰乙酸) to form a six-carbon compound(citrate,柠檬酸).

柠檬酸经过一系列的化学转换，失去两个羧基生成 CO_2。

循环过程中碳的去向。

● The citrate then goes through a series of chemical transformations, losing two carboxyl groups as CO_2. The carbons lost as CO_2 originate from what was oxaloacetate(草酰乙酸), not directly from acetyl-CoA. The carbons donated by acetyl-CoA become part of the oxaloacetate carbon backbone after the first turn of the citric acid cycle. Loss of the acetyl-CoA-donated carbons as CO_2 requires several turns of the citric acid cycle. However, because of the role of the citric acid cycle in anabolism, they may not be lost, since many TCA cycle intermediates are also used as precursors for the biosynthesis of other molecules.

大部分能量来源于循环过程中的氧化步骤。

● Most of the energy made available by the oxidative steps of the cycle is transferred as energy-rich electrons to NAD^+, forming NADH. For each acetyl group that enters the citric acid cycle, three molecules of NADH are produced.

● Electrons are also transferred to the electron acceptor Q (电子受体,辅酶 Q), forming QH_2.

每次循环最后，4 碳的草酰乙酸会重新生成。

● At the end of each cycle, the four-carbon oxaloacetate

has been regenerated, and the cycle continues.

2.5.2 Steps 步骤

Two carbon atoms are oxidized to CO_2, the energy from these reactions being transferred to other metabolic processes by GTP (or ATP), and as electrons in NADH and QH_2. The NADH generated in the TCA cycle may later donate its electrons in oxidative phosphorylation to drive ATP synthesis; $FADH_2$ is covalently (共价地) attached to succinate dehydrogenase(琥珀酸脱氢酶), an enzyme functioning both in the TCA cycle and the mitochondrial electron transport chain in oxidative phosphorylation. $FADH_2$, therefore, facilitates transfer of electrons to coenzyme Q, which is the final electron acceptor of the reaction catalyzed by the Succinate: ubiquinone oxidoreductase complex(琥珀酸:泛醌氧化还原酶复合物), also acting as an intermediate in the electron transport chain(电子传递链).

电子在三羧酸循环中传递的具体步骤。

琥珀酸脱氢酶在三羧酸循环和氧化磷酸化中均发挥作用。

Mitochondria in animals, including humans, possess two succinyl-CoA synthetases(琥珀酰辅酶A合成酶): one that produces GTP from GDP, and another that produces ATP from ADP. Plants have the type that produces ATP (ADP-forming succinyl-CoA synthetase). Several of the enzymes in the cycle may be loosely associated in a multienzyme protein complex(多酶蛋白质复合物) within the mitochondrial matrix.

动物和人的线粒体中有两种琥珀酰辅酶A合成酶,它们各自有不同的功能。

The citric acid cycle is continuously supplied with new carbon in the form of acetyl-CoA, entering at step 1 below.

	Substrates	Products	Enzyme	Reaction type	Comment
1	Oxaloacetate+ Acetyl CoA+H_2O	Citrate+ CoA—SH	Citrate synthase	Aldol condensation	irreversible, extends the 4C oxaloacetate to a 6C molecule
2	Citrate	*cis*-Aconitate +H_2O	Aconitase	Dehydration	reversible isomerization
3	*cis*-Aconitate +H_2O	Isocitrate		Hydration	
4	Isocitrate+NAD^+	Oxalosuccinate +NADH+H^+	Isocitrate dehydrogenase	Oxidation	generates NADH (equivalent of 2.5 ATP)
5	Oxalosuccinate	α-Ketoglutarate +CO_2		Decarboxylation	rate-limiting, irreversible stage, generates a 5C molecule

	Substrates	Products	Enzyme	Reaction type	Comment
6	α-Ketoglutarate+ NAD$^+$+ CoA—SH	Succinyl-CoA+ NADH+ H$^+$+CO$_2$	α-Ketoglutarate dehydrogenase	Oxidative decarboxylation	irreversible stage, generates NADH (equivalent of 2.5 ATP), regenerates the 4C chain (CoA excluded)
7	Succinyl-CoA+ GDP+Pi	Succinate+CoA- SH+GTP	Succinyl- CoA synthetase	substrate-level phosphorylation	or ADP→ATP instead of GDP→ GTP, generates 1 ATP or equivalent
8	Succinate+ ubiquinone (Q)	Fumarate+ ubiquinol (QH$_2$)	Succinate dehydrogenase	Oxidation	uses FAD as a prosthetic group (FAD →FADH$_2$ in the first step of the reaction) in the enzyme, generates the equivalent of 1.5 ATP
9	Fumarate+H$_2$O	L-Malate	Fumarase	H$_2$O addition	
10	L-Malate+ NAD$^+$	Oxaloacetate+ NADH+H$^+$	Malate dehydrogenase	Oxidation	reversible (in fact, equilibrium favors malate), generates NADH (equivalent of 2.5 ATP)

Reactants and products in the citric acid cycle.

Description	Reactants	Products
The sum of all reactions in the citric acid cycle is:	Acetyl-CoA+3NAD$^+$ +Q+GDP+P$_i$ +2H$_2$O	→CoA—SH+3NADH+3H$^+$ +QH$_2$+GTP+2CO$_2$
Combining the reactions occurring during the pyruvate oxidation with those occurring during the citric acid cycle, the following overall pyruvate oxidation reaction is obtained:	Pyruvate ion + 4 NAD$^+$+Q+GDP +P$_i$+2 H$_2$O	→4 NADH+4 H$^+$+QH$_2$+ GTP+3 CO$_2$
Combining the above reaction with the ones occurring in the course of glycolysis, the following overall glucose oxidation reaction (excluding reactions in the respiratory chain) is obtained:	Glucose + 10NAD$^+$ + 2Q + 2ADP + 2GDP+4P$_i$+2H$_2$O	→10NADH + 10H$^+$ + 2QH$_2$ +2ATP+2GTP+6CO$_2$

GTP 可以被二磷酸核苷酸激酶利用产生 ATP。

The GTP that is formed by GDP-forming succinyl-CoA synthetase may be utilized by nucleoside-diphosphate kinase(二磷酸核苷激酶) to form ATP (the catalyzed reaction is GTP+ ADP→GDP+ATP).

2.5.3 Products 产物

三羧酸循环的产物。

Products of the first turn of the cycle are: one GTP (or ATP), three NADH, one QH$_2$, two CO$_2$.

Because two acetyl-CoA molecules are produced from each glucose molecule, two cycles are required per glucose molecule. Therefore, at the end of two cycles, the products are: two GTP, six NADH, two QH_2, and four CO_2.

The total number of ATP obtained after complete oxidation of one glucose in glycolysis, citric acid cycle, and oxidative phosphorylation is estimated to be between 30 and 38. A recent assessment of the total ATP yield with the updated proton-to-ATP ratios provides an estimate of 29.85 ATP per glucose molecule.

2.5.4 Regulation 调节

The regulation of the TCA cycle is largely determined by substrate availability and product inhibition. NADH, a product of all dehydrogenases in the TCA cycle with the exception of succinate dehydrogenase, inhibits pyruvate dehydrogenase, isocitrate dehydrogenase(异柠檬酸脱氢酶), α-ketoglutarate dehydrogenase(α-酮戊二酸脱氢酶), and also citrate synthase (柠檬酸合酶). Acetyl-CoA inhibits pyruvate dehydrogenase, while succinyl-CoA inhibits α-ketoglutrate dehydrogenase and citrate synthase. When tested in vitro(体外) with TCA enzymes, ATP inhibits citrate synthase and α-ketoglutarate dehydrogenase; however, ATP levels do not change more than 10% in vivo between rest and vigorous exercise(剧烈运动). There is no known allosteric mechanism(别构机理) that can account for large changes in reaction rate from an allosteric effector(别构效应物) whose concentration changes less than 10%.

Calcium(钙) is used as a regulator. It activates pyruvate dehydrogenase, isocitrate dehydrogenase and α-ketoglutarate dehydrogenase. This increases the reaction rate of many of the steps in the cycle, and therefore increases flux throughout the pathway.

Citrate is used for feedback inhibition(反馈抑制), as it inhibits phosphofructokinase, an enzyme involved in glycolysis that catalyses formation of fructose 1, 6-bisphosphate, a precursor of pyruvate. This prevents a constant high rate of flux when there is an accumulation of citrate and a decrease in substrate for the enzyme.

Recent work has demonstrated an important link between intermediates of the citric acid cycle and the regulation of hypoxia-inducible factors(缺氧诱导因子，HIF). HIF plays a role in the regulation of oxygen homeostasis(氧气体内平衡), and is a transcription factor(转录因子) that targets angiogenesis(血管再生), vascular remodeling(血管重构), glucose utilization(葡糖利用), iron transport(铁运输) and apoptosis(细胞凋亡). HIF is synthesized constitutively, and hydroxylation(羟基化) of at least one of two critical proline residues(脯氨酸残基) mediates their interaction with the von Hippel Lindau E3 ubiquitin ligase complex(E3泛素连接酶复合物), which targets them for rapid degradation. This reaction is catalysed by prolyl 4-hydroxylases(脯氨酰羟化酶). Fumarate(延胡索酸) and succinate have been identified as potent inhibitors of prolyl hydroxylases, thus leading to the stabilization of HIF.

2.5.5 Major metabolic pathways converging on the TCA cycle 汇聚在 TCA 循环的主要代谢途径

Several catabolic pathways converge on the TCA cycle. Reactions that form intermediates of the TCA cycle in order to replenish(补充) them (especially during the scarcity of the intermediates) are called anaplerotic reactions(回补反应).

The citric acid cycle is the third step in carbohydrate catabolism(分解代谢)(the breakdown of sugars). Glycolysis breaks glucose (a six-carbon-molecule) down into pyruvate (a three-carbon molecule). In eukaryotes, pyruvate moves into the mitochondria. It is converted into acetyl-CoA by decarboxylation and enters the citric acid cycle.

In protein catabolism, proteins are broken down by proteases(蛋白酶) into their constituent amino acids. The carbon backbone of these amino acids can become a source of energy by being converted to acetyl-CoA and entering into the citric acid cycle.

In fat catabolism, triglycerides(甘油三酯) are hydrolyzed (水解) to break them into fatty acids and glycerol. In the liver the glycerol can be converted into glucose via dihydroxyacetone phosphate(二羟丙酮磷酸) and glyceraldehyde-3-phosphate by way of gluconeogenesis. In many tissues, especially heart tissue, fatty acids are broken down through a process known as beta oxidation, which results in acetyl-CoA, which can be used in the citric acid cycle. Beta oxidation of fatty acids with an odd number of methylene groups(亚甲基) produces propionyl CoA (丙酰辅酶A), which is then converted into succinyl-CoA and fed into the citric acid cycle.

脂肪分解代谢与TCAC的关联。

The total energy gained from the complete breakdown of one molecule of glucose by glycolysis, the citric acid cycle, and oxidative phosphorylation equals about 30 ATP molecules, in eukaryotes. The citric acid cycle is called an amphibolic pathway(两用代谢途径) because it participates in both catabolism and anabolism(合成代谢).

真核生物中，一分子葡萄糖通过糖酵解、三羧酸循环和氧化磷酸化降解，产生大约30个ATP分子。

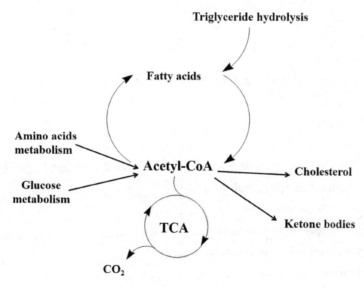

Interactive pathway map.

【Key words】

glycolysis 糖酵解	pyruvate carboxylase 丙酮酸羧化酶
alcoholic fermentation 乙醇发酵	hexokinase kinase 己糖激酶
lactic acid fermentation 乳酸发酵	phosphofructokinase (PFK) 磷酸果糖激酶
phosphorylation 磷酸化作用	α-ketoglutarate dehydrogenase α-酮戊二酸脱氢酶
enol phosphate 烯醇磷酸	pyruvate kinase 丙酮酸激酶
aerobic glycolysis 有氧酵解	citrate synthase 柠檬酸合酶
citric acid cycle 三羧酸循环	malate synthase 苹果酸合酶
anaplerotic reaction 回补反应	pyruvate dehydrogenase complex 丙酮酸脱氢酶复合体
gluconeogenesis 糖异生	isocitrate dehydrogenase 异柠檬酸脱氢酶
glucose-6-phosphatase 葡萄糖-6-磷酸	acetyl CoA 乙酰辅酶A

Questions

1. State the pathway of glycolysis.

2. What coenzymes are required by the pyruvate dehydrogenase complex? What are their roles? Before any oxidation can occur in the citric acid cycle, citrate must be isomerized into isocitrate. Why is this the case?

3. Briefly define the following terms:
   (a) Glycolysis
   (b) Aerobic oxidation of glucose
   (c) Anaerobic glycolysis
   (d) Citric acid cycle
   (e) Gluconeogenesis

4. Give the names of the regulatory enzymes for the following pathways:
   (a) Glycolysis
   (b) Gluconeogenesis
   (c) Citric acid cycle

5. Each of the following molecules is processed by glycolysis to lactate. How much ATP is generated from each molecule?
   (a) Glucose 6-phosphate
   (b) Dihydroxyacetone phosphate
   (c) Glyceraldehyde 3-phosphate
   (d) Fructose
   (e) Sucrose

References

[1] McKee Trudy, R McKee James. Biochemistry: An Introduction (Second Edition)[M]. New York: McGraw-Hill Companies, 1999.

[2] Nelson D L, Cox M M. Lehninger Principles of Biochemistry (Third Edition)[M]. Derbyshire: Worth Publishers, 2000.

[3] Ding W, Jia H T. Biochemistry (First Edition)[M]. Beijing: Higher Education Press, 2012.

[4] 郑集, 陈钧辉. 普通生物化学(第四版)[M]. 北京: 高等教育出版社, 2007.

[5] Berg J M, Tymoczko J L, Styer L. Biochemistry (Seventh Edition)[M]. New York: W. H. Freeman and Company, 2012.

Chapter 3　Lipid metabolism 脂质代谢

3.1　Fatty acid synthesis 脂肪酸合成

有关脂肪酸合成的酶。

Fatty acid synthesis is the creation of fatty acids from acetyl-CoA(乙酰辅酶 A) and malonyl-CoA(丙二酰辅酶 A) precursors through action of enzymes called fatty acid synthases. It is an important part of the lipogenesis process, which-together with glycolysis-stands behind creating fats from blood sugar in living organisms. Straight-chain fatty acids occur in two types: saturated and unsaturated.

3.1.1　Saturated Straight-Chain Fatty Acids 饱和直链脂肪酸

大肠杆菌中饱和脂肪酸的合成。

Synthesis of saturated fatty acids via Fatty Acid Synthase II in *E. coli*.

— 180 —

The diagrams presented show how fatty acids are synthesized in microorganisms and list the enzymes found in *Escherichia coli*. These reactions are performed by fatty acid synthase II (FASII，脂肪酸合成酶II), which contain multiple enzymes that generally act as one complex(复合体). FASII is present in prokaryotes, plants, fungi, parasites(寄生虫), as well as inmitochondria.

脂肪酸合成酶II在原核生物、植物、真菌、寄生虫、线粒体中都存在。

In animals, as well as yeast and some fungi, these same reactions occur on fatty acid synthase I (FASI), a large dimeric protein, which has all of the enzymatic activities required to create a fatty acid. FASI is less efficient than FASII, however, it allows for the formation of more molecules, including "medium-chain" fatty acids via early chain termination. Once a 16：0 carbon fatty acid has been formed it can undergo a number of modifications; particularly, by fatty acidsynthase III (FASIII), which uses 2 carbon molecules to elongate(延长) preformed fatty acids.

在动物、酵母和一些真菌中,由脂肪酸合成酶I催化同类反应。

(1) Regulation 调节

Acetyl-CoA is formed into malonyl-CoA by acetyl-CoA carboxylase(乙酰辅酶A羧化酶), at which point malonyl-CoA is destined to feed into the fatty acid synthesis pathway. Acetyl-CoA carboxylase is the point of regulation in saturated(饱和的) straight-chain fatty acid synthesis, by both phosphorylation(磷酸化作用) and allosteric regulation(别构调节). Regulation by phosphorylation occurs mostly in mammals(哺乳动物), while allosteric regulation occurs in most organisms. Allosteric control occurs as feed back inhibition(反馈抑制) by palmitol-CoA(软脂酰辅酶A) and activation(激活) by citrate. When there are high levels of palmital-CoA, the final product of saturated fatty acid synthesis, it allosterically inactivates acetyl-CoA carboxylase to prevent a build up of fatty acids in cells. Citrate acts to activate acetyl-CoA carboxylase under high levels, because high levels indicate there is enough acetyl-CoA to feed into the Krebs cycle and produce energy.

丙二酸单酰辅酶A由乙酰辅酶A经乙酰辅酶A羧化酶催化生成,然后丙二酸单酰辅酶A进入脂肪酸合成途径。

(2) De Novo Synthesisin Humans 人体中从头合成

In humans, fatty acids are predominantly formed in the liver and lactating mammary glands, and, to a lesser extent, the adipose tissue. Most acetyl-CoA is formed from pyruvate by pyruvate dehydrogenase in the mitochondria(线粒体). Acetyl-CoA produced in the mitochondria is condensed with oxaloacetate(草酰乙酸) by citrate synthase(柠檬酸合酶) to form citrate, which is then transported into the cytosol and broken down to yield acetyl-CoA and oxaloacetate by ATP citrate lyase. Oxaloacetate in the cytosol is reduced to malate by cytoplasmic malate dehydrogenase(苹果酸脱氢酶), and malate is transported back into the mitochondria to participate in the Citric acid cycle.

3.1.2 Desaturation 去饱和

Unsaturated(不饱和) fatty acids are essential components to prokaryotic and eukaryotic cell membranes. These fatty acids primarily function in maintaining membrane fluidity(流动性). They have also been associated with serving as signaling molecules in other processes such as cell differentiation(分化) and DNA replication. There are two pathways organisms use for desaturation: Aerobic and Anaerobic.

(1) Anaerobic Desaturation 厌氧去饱和

Many bacteria use the anaerobic(厌氧的) pathway for synthesizing unsaturated fatty acids. This pathway does not utilize(利用) oxygen and is dependent on enzymes to insert the double bond before elongation utilizing the normal fatty acid synthesis machinery. In *Escherichia coli*(大肠杆菌), this pathway is well understood.

● FabA is a β-hydroxydecanoyl-ACP dehydrase(β-羟基癸酰-ACP 脱水酶), it is specific for the 10-carbon saturated fatty acid synthesis intermediate(β-hydroxydecanoyl-ACP).

● FabA catalyzes the dehydration(脱水) of β-hydroxydecanoyl-ACP causing the release of water and insertion of the double bond between C_7 and C_8 counting from the methyl end. This creates the *trans-2-decanoyl* intermediate.

- The *trans-2-decanoyl* intermediate can either be shunted to the normal saturated fatty acid synthesis pathway by FabB, where the double bond will be hydrolyzed(水解的) and the final product will be a saturated fatty acid, or FabA will catalyze the isomerization into the *cis-3-decanoyl* intermediate.

- FabB is a β-ketoacyl-ACP synthase which elongates and channels intermediates into the mainstream(主流) fatty acid synthesis pathway. When FabB reacts with the *cis-decanoyl* intermediate, the final product after elongation will be an unsaturated fatty acid.

- The two main unsaturated fatty acids made are Palmitoleoyl-ACP (16:1ω7) and *cis*-vaccenoyl-ACP (18:1ω7). Most bacteria that undergo anaerobic desaturation contain homologues of FabA and FabB. Clostridia are the main exception; they have a novel enzyme, yet to be identified, that catalyzes the formation of the *cis* double bond.

Synthesis of unsaturated fatty acids via anaerobic desaturation.

(2) Aerobic Desaturation 有氧去饱和

Aerobic（需氧的） desaturation（去饱和） is the most widespread pathway for the synthesis of unsaturated fatty acids. It is utilized in all eukaryotes and some prokaryotes. This pathway utilizes desaturases（去饱和酶） to synthesize unsaturated fatty acids from full length saturated fatty acid substrates. All desaturases require oxygen and reducing equivalents acquired from the electron transport chain（电子传递链）. Desaturases are specific for the double bond they induce in the substrate. In *Bacillus subtilis*（枯草芽孢杆菌）, the desaturase, Δ^5 - Des（Δ^5 -去饱和酶系）, is specific for inducing a *cis*-double bond（反式双键） at the Δ^5 position. *Saccharomyces cerevisiae*（酿酒酵母） contains one desaturase, Ole1p, which induces the *cis*-double bond at Δ^9.

去饱和酶是一种混合功能氧化酶。

不饱和脂肪酸的有氧合成途径。

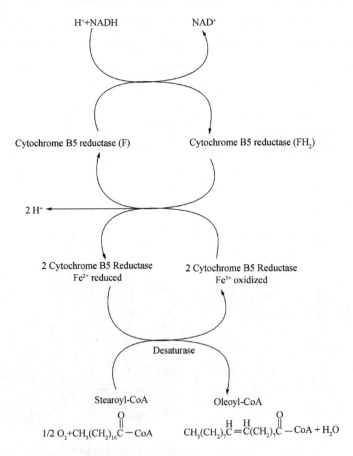

Synthesis of unsaturated fatty acids via aerobic desaturation.

3.1.3 Branched chain fatty acids 支链脂肪酸

Branched-chain fatty acids are usually saturated and are found in two distinct families: the iso-series and anteiso-series. It has been found that *Actinomycetales*(放线菌) contain unique branch chain fatty acid synthesis mechanisms(支链脂肪酸合成机制), including that which forms tuberculosteric acid.

(1) Branch-Chain Fatty Acid Synthesizing System 支链脂肪酸合成系统

The branched-chain fatty acid synthesizing system uses α-keto acids as primers(引物). This system is distinct from the branched-chain fatty acid synthetase which utilizes short-chain acyl-CoA esters as primers. α-keto acid primers are derived from the transamination(转氨作用) and decarboxylation(脱羧) of valine, leucine(异亮氨酸), and isoleucine to form 2-methylpropanyl-CoA(2-甲基丙酰辅酶 A), 3-methylbutyryl-CoA(3-甲基丁酰辅酶 A), and 2-methylbutyryl-CoA(2-甲基丁酰辅酶 A), respectively. 2-methylpropanyl-CoA primers derived from valine(缬氨酸) are elongated to produce even-numbered iso-series fatty acids such as 14-methyl-pentadecanoic (isopalmitic) acid(十五烷酸), and 3-methylbutyryl-CoA primers from leucine may be used to form odd numbered iso-series fatty acids such as 13-methyl-tetradecanoic acid. 2-Methylbutyryl-CoA primers from isoleucine are elongated to form anteiso-series fatty acids containing an odd number of carbon atoms such as 12-methyl tetradecanoic acid(十四酸). Decarboxylation of the primer precursors occurs through the branched-chain α-keto acid decarboxylase (BCKA) enzyme(α-酮酸脱羧酶). Elongation of the fatty acid follows the same biosynthetic pathway in *Escherichia coli* used to produce straight-chain fatty acids where malonyl-CoA is used as a chain extender(延伸器). The major end products are 12~17 carbon branched-chain fatty acids and their composition tends to be uniform and characteristic for many bacterial species.

支链脂肪酸的系列和合成机制。

支链脂肪酸合成系统的引物及合成过程。

(2) BCKA decarboxylase and relative activities of α‑keto acid substrates BCKA 脱羧酶和 α 酮酸底物的相对活性

The BCKA decarboxylase enzyme is composed of two subunits in a tetrameric(四聚物) structure（A_2B_2）and is essential for the synthesis of branched-chain fatty acids. It is responsible for the decarboxylation of α‑keto acids formed by the transamination(转氨作用) of valine, leucine, and isoleucine and produces the primers used for branched-chain fatty acid synthesis. The activity of this enzyme is much higher with branched‑chain α‑keto acid substrates than straight-chain substrates, and in Bacillus species its specificity is highest for the isoleucine-derived α‑keto‑β‑methylvaleric acid(α‑酮基‑β‑甲基戊酸), followed by α‑ketoisocaproate(α‑酮异己酸) and α‑ketoisovalerate(α‑酮异戊酸). The enzyme's high affinity toward branched‑chain α‑keto acids allows it to function as the primer donating system for branched-chain fatty acid synthetase.

BCKA decarboxylase and relative activities.

Substrate	BCKA activity	CO_2 Produced ($nmol \cdot min^{-1} \cdot mg^{-1}$)	K_m(μM)	V_{max} ($nmol \cdot min^{-1} \cdot mg^{-1}$)
L‑α‑keto‑β‑methyl‑valerate	100%	19.7	<1	17.8
α‑Ketoisovalerate	63%	12.4	<1	13.3
α‑Ketoisocaproate	38%	7.4	<1	5.6
Pyruvate	25%	4.9	51.1	15.2

(3) Factors affecting chain length and pattern distribution 影响链长度和分布模式的因素

α‑keto acid primers are used to produce branched-chain fatty acids that are generally between 12 and 17 carbons in length. The proportions(比例) of these branched-chain fatty acids tend to be uniform and consistent among a particular bacterial species but may be altered due to changes in malonyl-CoA (丙二酰-CoA) concentration, temperature, or heat-stable factors（HSF，热稳定因素）present. All of these factors may affect chain length, and HSFs have been demonstrated(被证明) to alter the specificity of BCKA decarboxylase for a particular α-keto acid substrate, thus shifting the ratio of branched-chain fatty acids produced. An increase in malonyl-CoA concentration has been shown to result in a larger proportion of C_{17} fatty acids produced, up until the optimal concentration (≈ 20 μM) of malonyl-CoA is reached. Decreased temperatures also tend to shift the fatty-acid distribution slightly toward C_{17} fatty-acids in *Bacillus* species.

(4) Branch-Chain Fatty Acid Synthase 支链脂肪酸合酶

This system functions similarly to the branch-chain fatty acid synthesizing system, however it uses short chain carboxylic acids as primers instead of alpha-keto acids(α-酮酸). This method is generally used by bacteria that do not have the ability to perform the branch-chain fatty acid system using alpha-keto primers. Typical short chain primers include isovalerate(异戊酸), isobutyrate(异丁酸) and 2-methyl butyrate(2-丁基甲酸). The acids needed for these primers are generally taken up from the environment; this is often seen in ruminal bacteria.

该系统的功能类似于支链脂肪酸合成系统,但它利用短链羧酸作为引物,而不是α-酮酸。

The overall reaction is:
Isobutyryl-CoA + 6malonyl-CoA + 12NADPH + 12H$^+$ = Isoplmitic acid + 6CO_2 + 12NADP + 5H_2O + 7CoA

The difference between (straight-chain) fatty acid synthase and branch-chain fatty acid synthase is substrate specificity of the enzyme that catalyzes the reaction of acyl-CoA to acyl-ACP.

直链脂肪酸合酶和支链脂肪酸合酶之间的区别是催化酰基-CoA 生成脂酰-ACP反应的酶的底物专一性。

3.1.4 Lipogenesis 脂肪生成

Lipogenesis(脂肪生成) is the process by which acetyl-CoA is converted to fats. The former is an intermediate(媒介) stage in metabolism of simple sugars, such as glucose, a source of energy of living organisms. Through lipogenesis, the energy can be efficiently stored in the form of fats. Lipogenesis encompasses(包含) the processes of fatty acid synthesis and subsequent triglyceride synthesis (where fatty acids are

脂肪生成,可有效地存储能量。

esterified(酯化) with glycerol to form fats). The products are secreted from the liver in the form of very-low-density lipoproteins(VLDL, 低密度脂蛋白).

(1) Fatty acid synthesis 脂肪酸合成

脂肪酸合成的场所。

Fatty acids synthesis starts with acetyl-CoA and builds up by the addition of two carbon units. The synthesis occurs in the cytoplasm(细胞质) in contrast to the degradation (oxidation), which occurs in the mitochondria(线粒体). Many of the enzymes for the fatty acid synthesis are organized into a multienzyme complex(多酶复合物) called fatty acid synthetase.

(2) Control and regulation 调控

胰岛素刺激脂肪合成的两种方式。

Insulin(胰岛素) is an indicator of the blood sugar level of the body, as its concentration increases proportionally with blood sugar levels. Thus, a large insulin level is associated with the fed state(储备状态). As one might expect, therefore, it increases the rate of storage pathways, such as lipogenesis. Insulin stimulates lipogenesis in two main ways: The enzymes pyruvate dehydrogenase (PDH), which forms acetyl-CoA, and acetyl-CoA carboxylase (ACC), which forms malonyl-CoA, are obvious control points. These are activated by insulin. So a high insulin level leads to an overall increase in the levels of malonyl-CoA, which is the substrate required for fatty acids synthesis.

(3) PDH dephosphorylation 丙酮酸脱氢酶脱磷酸作用

高水平的胰岛素可提高丙酮酸脱氢酶脱磷酸作用。

Pyruvate dehydrogenase(丙酮酸脱氢酶) dephosphorylation is increased with the release of insulin. The dephosphorylated form is more active.

As insulin binds to cellular surface transmembrane receptors that intracellularly activate the adenylate cyclase enzyme that catalyze cAMP (cyclic AMP, 环腺苷酸) production from ATP. The increased intracellular(胞内的) cAMP, acts as a second messenger(第二信使), in response to the insulin binding. cAMP activates protein kinase enzyme(蛋白激酶) that in turn activates phosporylase enzyme that phosphorylates(磷酸

化）and in doing so activates a number of different intracellular enzymes such as the pyruvate dehydrogenase that dehydrates pyruvate to form AcCoA（乙酰辅酶 A）. So, an extracellular hormone, insulin, can in multistep activation（cascade, 级联）activate an enzyme in the cellular matrix.

This mechanism leads to the increased rate of catalysis of this enzyme, so increases the levels of acetyl-CoA. Increased levels of acetyl-CoA will increase the flux through not only the fat synthesis pathway but also the citric acid cycle.

3.1.5 Acetyl-CoA carboxylase 乙酰辅酶 A 羧化酶

Insulin affects ACC in a similar way to PDH. It leads to its phosphorylation, activation. But as the fatty acid is synthesised it will have a negative feedback on its own enzyme, by dephosporylation, deactivation of protein kinaseand in so deactivation of ACC. Glucagon has an antagonistic effect（拮抗作用）and increases phosphorylation, deactivation, thereby inhibiting ACC and slowing fat synthesis.

胰岛素影响乙酰辅酶 A 羧化酶的方式与影响丙酮酸脱氢酶的方式相似。

Affecting ACC affects the rate of acetyl-CoA conversion to malonyl-CoA. Increased malonyl-CoA level pushes the equilibrium over to increase production of fatty acids through biosynthesis.

影响乙酰辅酶 A 羧化酶，则会影响乙酰辅酶 A 转化成丙二酰辅酶 A 的速率。

AMP and ATP concentrations of the cell act as a measure of the ATP needs of a cell and as ATP levels get low it activates the ATP synthease which in turn phosphorylates ACC. When ATP is depleted, there is a rise in 5'AMP. This rise activates AMP-activated protein kinase, which phosphorylates ACC, thereby inhibits fat synthesis. This is a useful way to ensure that glucose is not diverted down a storage pathway in times when energy levels are low. ACC is also activated by citrate. This means that, when there is abundant acetyl-CoA in the cell cytoplasm for fatsynthesis, it proceeds at an appropriate rate.

AMP 和 ATP 的浓度也会影响 ACC，当 ATP 含量低，它激活 ATP 合酶，进而影响 ACC 的磷酸化。

Note: Research now shows that glucose metabolism（exact

metabolite to be determined), aside from insulin's influence on lipogenic enzyme(脂肪酶) genes, can induce the gene products for liver's pyruvate kinas(丙酮酸激酶), acetyl-CoA carboxylase (乙酰辅酶A羧化酶), and fatty acid synthase(脂肪酸合酶). These genes are induced by the transcription factors(转录因子) ChREBP/Mlx via high blood glucose levels and presently unknown signaling events(信号活动). Insulin induction is due to SREBP-1c(固醇调节元件结合蛋白-1c), which is also involved in cholesterol metabolism(胆固醇代谢).

Acetyl-CoA carboxylase（ACC）is a biotin（生物素）-dependent enzyme that catalyzes the irreversible carboxylation of acetyl-CoA to produce malonyl-CoA through its two catalytic activities, biotin carboxylase（BC，生物素羧化酶）and carboxyl transferase（CT）（羧基转移酶）. ACC is a multi-subunit enzyme in most prokaryotes and in the chloroplasts of most plant sand algae, whereas it is a large, multi-domain enzyme in the endoplasmic（内质网）reticulum of most eukaryotes. The most important function of ACC is to provide the malonyl-CoA substrate for the biosynthesis of fatty acids. The activity of ACC can be controlled at the transcriptional level(转录水平) as well as by small molecule modulators（调节器）and covalent modification. The human genome contains the genes for two different ACCs — ACACA and ACACB.

（1）Structure 结构

Prokaryotes and plants have multi-subunit ACCs composed of several polypeptides encoded by distinct genes. Biotin carboxylase（BC）activity, biotin carboxyl carrier protein（BCCP）, and carboxyl transferase（CT）activity are each contained on a different subunit. The stoichiometry(化学计量学) of these subunits in the ACC holoenzyme differs amongst organisms. Humans and most eukaryotes have evolved an ACC with CT and BC catalytic domains and biotin carboxyl carrier domains on a single polypeptide. ACC functional regions, starting from the N-terminus to C-terminus are the biotin carboxylase（BC）, biotin binding（BB）, carboxyltransferase（CT）, and ATP-binding（AB）. AB lies within BC. Biotin is covalently attached through an amide bond(酰胺基) to the long side chain of a lysine reside in BB. As BB is between BC and CT regions, biotin can easily translocate to both of the active sites where it is required.

In mammals where two isoforms(亚型) of ACC are expressed, the main structural difference between these isoforms is the extended ACC2 N-terminus containing a mitochondria(线粒体) targeting sequence.

哺乳动物体内 ACC 两个亚型的结构差别。

(2) Mechanism 机制

The overall reaction of ACAC(A, B) proceeds by a two-step mechanism. The first reaction is carried out by BC and involves the ATP-dependent carboxylation of biotin with bicarbonate(碳酸氢盐) serving as the source of CO_2. The carboxyl group is transferred from biotin to acetyl CoA to form malonyl CoA in the second reaction, which is catalyzed by CT.

ACAC 反应可分为两步，第一步用碳酸氢盐作为碳源，耗能；第二步用 CT 催化。

The reaction mechanism of ACAC(A, B).

In the context of the active site, the reaction proceeds with extensive interaction(相互作用) of the residues Glu296 and positively charged Arg338 and Arg292 with the substrates. Two Mg^{2+} are coordinated(协调的) by the phosphate groups on the ATP, and are required for ATP binding to the enzyme. Bicarbonate is deprotonated by Glu296, although in solution, this proton transfer is unlikely as the pK_a of bicarbonate is 10.3. The enzyme apparently manipulates pK_a to facilitate the deprotonation(去质子化) of bicarbonate. The pK_a of bicarbonate is decreased by its interaction with positively

参加反应的底物、酶、反应环境的介绍。

charged side chains of Arg338 and Arg292. Furthermore, Glu296 interacts with the side chain of Glu211, an interaction that has been shown to cause an increase in the apparent pK_a. Following deprotonation of bicarbonate, the oxygen of the bicarbonate acts as a nucleophile(亲核试剂) and attacks the gamma phosphate on ATP. The carboxy phosphate intermediate quickly decomposes(分解) to CO_2 and PO_4^{3-}. The PO_4^{3-} deprotonates biotin, creating an enolate(烯醇化物), stabilized by Arg338, that subsequently attacks CO_2 resulting in the production of carboxybiotin. The carboxybiotin translocates to the carboxy transferase(CT,羧基转移酶) active site, where the carboxyl group is transferred to acetyl-CoA. In contrast to the BC domain, little is known about the reaction mechanism of CT. A proposed mechanism is the release of carbon dioxide from biotin, which subsequently abstracts a proton from the methyl group from acetyl CoA carboxylase. The resulting enolate(烯醇) attacks CO_2 to form malonyl CoA. In a competing mechanism, proton abstraction is concerted with the attack of acetyl CoA.

(3) Function 功能

The function of ACC is to regulate the metabolism of fatty acids. When the enzyme is active, the product, malonyl-CoA is produced which is a building block for new fatty acids and can inhibit the transfer of the fatty acyl group from acyl CoA to carnitine(肉毒碱) with carnitine acyltransferase(酰基转移酶), which inhibits the beta-oxidation of fatty acids in the mitochondria.

In mammals, two main isoforms of ACC are expressed, ACC1 and ACC2, which differ in both tissue distribution and function. ACC1 is found in the cytoplasm of all cells but is enriched in lipogenic(脂肪生成) tissue, such as adipose tissue and lactating mammary glands(乳腺), where fatty acid synthesis is important. In oxidative tissues, such as the skeletal muscle(骨骼肌) and the heart, the ratio of ACC2 expressed is higher. ACC1 and ACC2 are both highly expressed in the liver where both fatty acid oxidation and synthesis is important. The differences in tissue distribution indicate that ACC1 maintains regulation of fatty acid synthesis whereas ACC2 mainly regulates fatty acid oxidation.

(4) Regulation 调节

The regulation of mammalian(哺乳类) ACC is complex, in order to control two distinct pools of malonyl CoA that direct either the inhibition of beta oxidation or the activation of lipid biosynthesis.

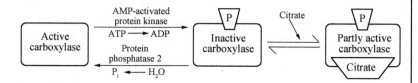

Control of Acetyl CoA Carboxylase. The AMP regulated kinase triggers the phosphorylation of the enzyme（thus inactivating it）and the phosphatase enzyme removes the phosphate group.

Mammalian ACC1 and ACC2 are regulated transcriptionally by multiple promoters which mediate ACC abundance in response to the cells nutritional status(营养状况). Activation of gene expression through different promoters results in alternative splicing(可变剪接); however, the physiological significance of specific ACC isozymes remains unclear. The sensitivity to nutritional status results from the control of these promoters by transcription factors such as SREBP1c(固醇调节元件结合蛋白-1c), controlled by insulin at the transcriptional level, and ChREBP, which increases in expression with high carbohydrates diets.

哺乳动物体内 ACC 的多转录调节。

Through a feedforward(前馈) loop, citrate(柠檬酸盐) allosterically activates ACC. Citrate may increase ACC polymerization to increases enzymatic activity; however, it is unclear if polymerization is citrate's main mechanism of increasing ACC activity or if polymerization is an artifact of in vitro experiments(体外实验). Other allosteric activators include glutamate(谷氨酸) and other dicarboxylic acids(二羧酸). Long and short chain fatty acyl CoAs are negative

柠檬酸激活 ACC 的机制。

feedback(负反馈) inhibitors of ACC.

Phosphorylation can result when the hormones(激素) glucagon(胰高血糖素) or epinephrine(肾上腺素) bind to cell surface receptors, but the main cause of phosphorylation is due to a rise in AMP levels when the energy status of the cell is low, leading to the activation of the AMP-activated protein kinase (AMPK). AMPK is the main kinase regulator of ACC, able to phosphorylate a number of serine(丝氨酸) residues on both isoforms of ACC. On ACC1, AMPK phosphorylates Ser79, Ser1200, and Ser1215. On ACC2, AMPK phosphorylates Ser218. Protein kinase A also has the ability to phosphorylate ACC, with a much greater ability to phosphorylate ACC2 than ACC1. However, the physiological significance of protein kinase A in the regulation of ACC is currently unknown. Researchers hypothesize there are other ACC kinases important to its regulation as there are many other possible phosphorylation sites on ACC.

When insulin binds to its receptors on the cellular membrane, it activates a phosphatase to dephosphorylate the enzyme; thereby removing the inhibitory effect.

(5) Clinical implications 临床意义

At the juncture(连接) of lipid synthesis and oxidation pathways, ACC presents many clinical possibilities for the production of novel antibiotics(抗生素) and the development of new therapies(治疗方法) for diabetes(糖尿病), obesity(肥胖) and other manifestations of metabolic syndrome(代谢综合征). Researchers aim to take advantage of structural differences between bacterial and human ACCs to create antibiotics specific to the bacterial ACC, in efforts to minimize side effects to patients. Promising results for the usefulness of an ACC inhibitor include the finding that ACC2 −/− mice (mice with no expression of ACC2) have continuous fatty acid oxidation, reduced body fat mass, and reduced body weight despite an increase in food consumption. ACC2 −/− mice are also protected from diabetes. It should be noted that mutant mice lacking ACC1 are embryonically lethal. However, it is unknown whether drugs targeting ACCs in humans must be specific for ACC2.

3.2 Fatty acid degradation 脂肪酸降解

Fatty acid degradation is the process in which fatty acids are broken down into their metabolites, in the end generating acetyl-CoA, the entry molecule for the citric acid cycle, the main energy supply of animals. It includes three major steps:
- Lipolysis of and release from adipose tissue;
- Activation and transport into mitochondria;
- β-oxidation.

3.2.1 Lipolysis and release 脂肪分解和释放

Initially in the process of degradation, fatty acids are stored in fat cells (adipocytes, 脂肪细胞). The breakdown of this fat is known as lipolysis. The products of lipolysis, free fatty acids, are released into the bloodstream and circulate throughout the body.

(1) **Activation and transport into mitochondria** 激活和运输到线粒体

Fatty acids must be activated before they can be carried into the mitochondria, where fatty acid oxidation occurs. This process occurs in two steps catalyzed by the enzyme fatty acyl-CoA synthetase.

(2) **Formation of an activated thioester bond** 活化硫酯键的形成

The enzyme first catalyzes nucleophilic attack on the α-phosphate of ATP to form pyrophosphate and an acyl chain linked to AMP. The next step is formation of an activated thioester bond between the fatty acyl chain and coenzyme A.

The formula(公式) for the above is:

$$RCOO^- + CoA + ATP + H_2O \longrightarrow RCO-CoA + AMP + PP_i + 2H^+$$

This two-step reaction is freely reversible and its equilibrium(均衡) lies near 1. To drive the reaction forward, the reaction is coupled to a strongly exergonic(放能) hydrolysis reaction: the enzyme inorganic pyrophosphatase(无机焦磷酸酶) cleaves the pyrophosphate liberated from ATP to two phosphate ions. Thus the net reaction becomes:

$$RCOO^- + CoA + ATP + H_2O \longrightarrow RCO-CoA + AMP + 2P_i + 2H^+$$

(3) Transport into the mitochondrial matrix 运输到线粒体基质

The inner mitochondrial membrane is impermeable(不渗透性的) to fatty acids and a specialized carnitine carrier system operates to transport activated fatty acids from cytosol to mitochondria.

Once activated, the acyl CoA is transported into the mitochondrial matrix. This occurs via a series of similar steps:

1. Acyl CoA is conjugated(共轭的) to carnitine by carnitine acyltransferase(酰基转移酶) (palmitoyltransferase,棕榈酰转移酶) I located on the outer mitochondrial membrane.
2. Acyl carnitine(肉碱) is shuttled(穿梭) inside by a translocase(移位酶).

3. Acyl carnitine (such as Palmitoyl carnitine) is converted to acyl CoA by carnitine acyl transferase (palmitoyl transferase) II located on the inner mitochondrial membrane. The liberated carnitine returns to the cytosol.

It is important to note that carnitine acyltransferase I undergoes allosteric inhibition as a result of malonyl-CoA, an intermediate in fatty acid biosynthesis, in order to prevent futile cycling (无效循环) between beta-oxidation and fatty acid synthesis.

3.2.2 β- oxidation β-氧化

Once inside the mitochondria, the β- oxidation of fatty acids occurs via five recurring steps:

1. Activation by ATP,
2. Oxidation by FAD,
3. Hydration(水合),
4. Oxidation by NAD^+,
5. Thiolysis(硫解),

The final product is acetyl-CoA, the entry molecule for the citric acid cycle.

线粒体内部β氧化的步骤。

(1) β oxidation β-氧化

β oxidation is the process by which fatty acids, in the form of Acyl-CoA molecules, are broken down in mitochondria and/or in peroxisomes(过氧化物酶体) to generate Acetyl-CoA, the entry molecule for the Citric Acid cycle(柠檬酸循环). The β oxidation of fatty acids involve three stages:

1. Activation of fatty acids in the cytosol.
2. Transport of fatty acids into mitochondria (carnitine shuttle).
3. β oxidation proper in the mitochondrial matrix.

Fatty acids are oxidized by most of the tissues in the body. However, the brain can hardly utilize fatty acids for energy requirements, while erythrocytes(红细胞) and adrenal medulla (肾上腺髓质) cannot use them at all.

微课：脂肪酸β氧化

脂肪酸β氧化的三个阶段。

脂肪酸 β 氧化示意图。

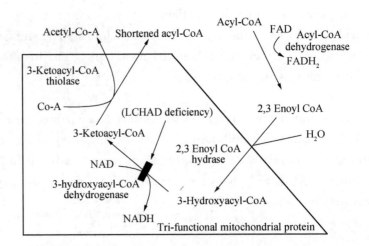

Schematic demonstrating mitochondrial fatty acid β-oxidation and effects of long-chain 3-hydroxyacyl-coenzyme A dehydrogenase deficiency, LCHAD deficiency.

（2）Activation of fatty acids 脂肪酸活化

Free fatty acids can penetrate(渗透) the plasma membrane due to their poor water solubility and high fat solubility. Once in the cytosol, activation of the fatty acid is catalyzed by long fatty acyl CoA synthetase. A fatty acid reacts with ATP to give a fatty acyl adenylate（腺苷酸）, plus inorganic（无机）pyrophosphate, which then reacts with free coenzyme A to give a fatty acyl-CoA ester plus AMP. The fatty acyl-CoA is then reacted with carnitine(肉毒素) to form acylcarnitine, which is transported across the inner mitochondrial membrane by a monosodium glutamate(谷氨酸).

（3）Four recurring steps 四个步骤重复

每次 β 氧化都释放一个两碳的乙酰辅酶 A。
每次循环都会生成一个乙酰辅酶 A、一个 NADH 和一个 FADH$_2$。

Once inside the mitochondria, each cycle of β-oxidation, liberating a two carbon unit-acetyl CoA, occurs in a sequence of four reactions; This process continues until the entire chain is cleaved into acetyl CoA units. The final cycle produces two separate acetyl CoAs, instead of one acyl CoA and one acetyl CoA. For every cycle, the Acyl CoA unit is shortened by two carbon atoms. Concomitantly, one molecule of FADH$_2$, NADH and acetyl CoA are formed.

(4) β- Oxidation of unsaturated fatty acids 不饱和脂肪酸的氧化

β- Oxidation of unsaturated fatty acids poses a problem since the location of a cis bond can prevent the formation of a $trans-\Delta^2$ bond. These situations are handled by an additional two enzymes.

不饱和脂肪酸的β氧化。

Whatever the conformation(构造) of the hydrocarbon chain, β-oxidation occurs normally until the acyl CoA (because of the presence of a double bond) is not an appropriate substrate for acyl CoA dehydrogenase, or enoyl CoA hydratase:

需要烯酯酰 CoA 异构酶的参与

- If the acyl CoA contains a $cis-\Delta^3$ bond, then $cis-\Delta^3$-enoyl CoA isomerase(顺-Δ^3-烯酯酰 CoA 异构酶) will convert the bond to a $trans-\Delta^2$ bond, which is a regular substrate.
- If the acyl CoA(脂酰 CoA) contains a $cis-\Delta^4$ double bond, then its dehydrogenation yields a 2,4-dienoyl intermediate(中间的), which is not a substrate for enoyl CoA hydratase(水合酶). However, the enzyme 2,4 Dienoyl CoA reductase reduces the intermediate, using NADPH, into $trans-\Delta^3-$ enoyl CoA. As in the above case, this compound is converted into a suitable intermediate by 3,2-Enoyl CoA isomerase.

如含有顺-Δ^4 不饱和键，则脱氢生成 2,4-二烯酰的中间产物。

To summarize:
- Odd-numbered(奇数的) double bonds are handled by the isomerase.
- Even-numbered(偶数的) double bonds by the reductase (还原酶). (which creates an odd-numbered double bond)

(5) β- Oxidation of odd-numbered chains 奇数链的氧化

In general, fatty acids with an odd number of carbon are found in the lipids of plants and some marine organisms(海洋生物). Many ruminant animals form large amount of 3-carbon propionate(丙酸) during fermentation(发酵) of carbohydrate in rumen. Chains with an odd-number of carbons are oxidized in the same manner as even-numbered chains, but the final products are propionyl-CoA and succinyl-CoA.

奇数碳脂肪酸链的β氧化。

Propionyl-CoA is first carboxylated using a bicarbonate ion into D-stereoisomer（立体异构） of methylmalonyl-CoA, in a reaction that involves a biotin co-factor（辅助因子）, ATP, and the enzyme propionyl-CoA carboxylase. The bicarbonate ion's carbon is added to the middle carbon of propionyl-CoA, forming a D-methylmalonyl（甲基丙二酰）-CoA. However, the D conformation is enzymatically（酶促的） converted into the L conformation by methylmalonyl-CoA epimerase（差向异构酶）, then it undergoes intramolecular rearrangement, which is catalyzed by methylmalonyl-CoA mutase (requiring B_{12} as a coenzyme) to form succinyl-CoA（琥珀酰辅酶A）. The succinyl-CoA formed can then enter the citric acid cycle.

Because it cannot be completely metabolized in the citric acid cycle, the products of its partial reaction must be removed in a process called cataplerosis. This allows regeneration of the citric acid cycle intermediates（中间物）, possibly an important process in certain metabolic diseases.

（6）Oxidation in peroxisomes 过氧化物酶体中的氧化

Fatty acid oxidation also occurs in peroxisomes, when the fatty acid chains are too long to be handled by the mitochondria. However, the oxidation ceases at octanyl CoA. It is believed that very long chain (greater than C-22) fatty acids undergo initial oxidation in peroxisomes which is followed by mitochondrial oxidation. One significant difference is that oxidation in peroxisomes is not coupled to ATP synthesis. Instead, the high-potential electrons are transferred to O_2, which yields H_2O_2. The enzyme catalase, found exclusively（唯一的） in peroxisomes, converts the hydrogen peroxide into water and oxygen.

Peroxisomal β-oxidation also requires enzymes specific to the peroxisome and to very long fatty acids. There are three key differences between the enzymes used for mitochondrial and peroxisomal β-oxidation：

1. β-oxidation in the peroxisome requires the use of a

peroxisomal carnitine acyltransferase(肉毒碱酰基转移酶)(instead of carnitine acyltransferase I and II used by the mitochondria) for transport of the activated acyl group into the peroxisome.

2. The first oxidation step in the peroxisome is catalyzed by the enzyme acyl-CoA oxidase.

3. The β- ketothiolase(硫解酶) used in peroxisomal β- oxidation has an altered substrate specificity, different from the mitochondrial β- ketothiolase.

Peroxisomal oxidation is induced by high-fat diet and administration of hypolipidemic drugs like clofibrate(安妥明).

(7) Energy yield 能量输出

The ATP yield for every oxidation cycle is 14 ATP (according to the P/O ratio), broken down as follows:

每次氧化产能 14 个 ATP。

Source	ATP	Total
1 FADH$_2$	×1.5 ATP	=1.5 ATP (some sources say 2 ATP)
1 NADH	×2.5 ATP	=2.5 ATP (some sources say 3 ATP)
1 acetyl CoA	×10 ATP	=10 ATP (some sources say 12 ATP)
TOTAL		=14 ATP

For an even-numbered saturated fat (C_{2n}), n－1 oxidations are necessary, and the final process yields an additional acetyl CoA. In addition, two equivalents(当量) of ATP are lost during the activation of the fatty acid. Therefore, the total ATP yield can be stated as:

$$(n-1)\times 14+10-2=\text{total ATP}$$

For instance, the ATP yield of palmitate (C_{16}, $n=8$) is:

$$(8-1)\times 14+10-2=106 \text{ ATP}$$

Represented in table form:

Source	ATP	Total
7 FADH$_2$	× 1.5 ATP	=10.5 ATP
7 NADH	×2.5 ATP	=17.5 ATP

（续表）

Source	ATP	Total
8 acetyl CoA	×10 ATP	=80 ATP
Activation		=−2 ATP
NET		=106 ATP

For sources that use the larger ATP production numbers described above, the total would be 129 ATP=(8−1)× 17+12−2 equivalents per palmitate. β-oxidation of unsaturated fatty acids changes the ATP yield due to the requirement of two possible additional enzymes.

3.3 Ketone bodies 酮体

酮体的组成及产生条件。

Ketone bodies are three water-soluble molecules (acetoacetate 乙酰乙酸, β-hydroxybutyrate β-羟丁酸, and their spontaneous breakdown product, acetone 丙酮) that are produced by the liver from fatty acids during periods of low food intake (fasting 禁食), carbohydrate restrictive diets, starvation, prolonged intense exercise, or in untreated (or inadequately treated) type 1 diabetes mellitus(糖尿病). These ketone bodies are readily picked up by the extra-hepatic tissues, and converted into acetyl-CoA which then enters the citric acid cycle and is oxidized in the mitochondria for energy. In the brain, ketone bodies are also used to make acetyl-CoA into long-chain fatty acids. The latter cannot be obtained from the blood, because they cannot pass through the blood-brain barrier (血脑屏障).

酮体可以被肝外组织作为能源物质使用。

脂肪酸 β-氧化产生的乙酰辅酶 A 通常情况下进入三羧酸循环彻底氧化。

Production

The acetyl-CoA produced by β-oxidation of fatty acids enters the citric acid cycle in the mitochondrion by combining with oxaloacetate(草酰乙酸) to form citrate(柠檬酸). This results in the complete combustion of the acetyl group of acetyl-CoA to CO_2 and water. The energy released in this process is captured in the form of 1 GTP and 11 ATP molecules per acetyl group (or acetic acid molecule) oxidized. This is the fate of acetyl-CoA wherever β- oxidation of fatty acids occurs, except under certain circumstances in the liver.

In the liver oxaloacetate(草酰乙酸) is wholly or partially diverted into the gluconeogenic pathway(糖异生途径) during fasting, starvation, a low carbohydrate diet, prolonged strenuous exercise, and in uncontrolled type 1 diabetes mellitus. Under these circumstances oxaloacetate is hydrogenated to malate(苹果酸) which is then removed from the mitochondrion to be converted into glucose in the cytoplasm of the liver cells, from where it is released into the blood. In the liver, therefore, oxaloacetate is unavailable for condensation with acetyl-CoA when significant gluconeogenesis(糖异生) has been stimulated by low (or absent) insulin(胰岛素) and high glucagon(胰高血糖素) concentrations in the blood. Under these circumstances acetyl-CoA is diverted to the formation of acetoacetate(乙酰乙酸) and β-hydroxybutyrate(β-羟丁酸). Acetoacetate, β-hydroxybutyrate, and their spontaneous breakdown product, acetone(丙酮), are frequently, but confusingly, known as ketone bodies (as they are not "bodies" at all, but water-soluble chemical substances).

在禁食、饥饿、低糖水化合物饮食、长时间剧烈运动或没有控制的1型糖尿病的情况下，糖异生作用使得肝脏的草酰乙酸被全部或部分耗竭，脂肪酸β氧化产生的乙酰辅酶A转向生成乙酰乙酸和β-羟丁酸，后两者自然分解产生丙酮。

乙酰乙酸、β-羟丁酸与丙酮三者合称酮体，均是水溶性物质。

Acetoacetate(乙酰乙酸) has a highly characteristic smell, for the people who can detect this smell, which occurs in the breath and urine(尿液) during ketosis. On the other hand, most people can smell acetone(丙酮), whose "sweet & fruity" odor(气味) also characterizes the breath of persons in ketosis (酮症) or, especially, ketoacidosis(酮症酸中毒).

乙酰乙酸具有特殊的气味，可以在呼气及尿液中被发现。
大部分人可以闻出酮症或酮症酸中毒病人呼气中的具有甜水果味的丙酮。

【Key words】

adipocyte 脂肪细胞	odd-numbered chains 奇数链
lipase 脂肪酶	glycerol 甘油
lipolysis 脂肪分解	degradation 降解
lipogenesis 脂肪生成	thiolysis 硫解

(续表)

fat mobilization 脂肪动员	dehydrogenation 脱氢
β-oxidation β氧化	hydration 水合作用
carnitine 肉碱	de novo synthesis 从头合成
carnitine acyltransferase (CAT) 肉碱脂酰基转移酶	peroxisomes 过氧化物酶体
acetoacetate 乙酰乙酸	acetone 丙酮
ketone bodies 酮体	β-hydroxybutyrate β-羟丁酸

Questions

1. Write a balanced equation for the conversion of stearate into acetoacetate.
2. Place the following list of reactions or relevant locations in the β-oxidation of fatty acids in the proper order.
   (a) Reaction with carnitine
   (b) Fatty acid in the cytoplasm
   (c) Activation of fatty acid by joining to CoA
   (d) Hydration
   (e) NAD^+-linked oxidation
   (f) Thiolysis
   (g) Acyl CoA in mitochondrion
   (h) FAD-linked oxidation
3. Suggest how fatty acids with odd numbers of carbons are synthesized.

References

[1] McKee Trudy, R McKee James. Biochemistry: An Introduction (Second Edition)[M]. New York: McGraw-Hill Companies, 1999.

[2] Nelson D L, Cox M M. Lehninger Principles of Biochemistry (Third Edition) [M]. Derbyshire: Worth Publishers, 2000.

[3] Ding W, Jia H T. Biochemistry (First Edition) [M]. Beijing: Higher Education Press, 2012.

[4] 郑集,陈钧辉. 普通生物化学(第四版)[M]. 北京:高等教育出版社,2007.

[5] Berg J M, Tymoczko J L, Styer L. Biochemistry (Seventh Edition)[M]. New York: W. H. Freeman and Company, 2012.

Part II Metabolism 新陈代谢

Chapter 4 Nitrogen metabolism 氮代谢

4.1 Nitrogen fixation 固氮作用

Nitrogen fixation(固氮作用) is the natural process, either biological or abiotic, by which nitrogen (N_2) in the atmosphere is converted into ammonia (NH_3). This process is essential for life, because fixed nitrogen is required to biosynthesize(生物合成) the basic building blocks of life, e. g., nucleotides(核苷酸) for DNA and RNA and amino acids for proteins. Nitrogen fixation also refers to other biological conversions of nitrogen, such as its conversion to nitrogen dioxide.

固氮作用是自然过程,在生物或非生物作用下,N_2可转换为NH_3。

Microorganisms(微生物) that fix nitrogen are bacteria called diazotrophs(固氮生物). Some higher plants, and some animals (termites 白蚁), have formed associations (symbioses) with diazotrophs. Nitrogen fixation also occurs as a result of non-biological processes. These include lightning, industrially through the Haber-Bosch Process, and combustion. Biological nitrogen fixation was discovered by the Dutch microbiologist Martinus Beijerinck.

固氮菌的介绍。

生物固氮是由荷兰微生物学家 Martinus Beijerinck 发现的。

Biological nitrogen fixation 生物氮固定

Biological nitrogen fixation (BNF) occurs when atmospheric nitrogen is converted to ammonia(氨) by an enzyme called nitrogenase(固氮酶). The reaction for BNF is:

$$N_2 + 8H^+ + 8e^- \longrightarrow 2NH_3 + H_2$$

The process is coupled to the hydrolysis(水解) of 16 equivalents(等价物) of ATP and is accompanied by the co-

在自生固氮菌中,固氮酶生成的氨,通过谷氨酰胺合成酶/谷氨酸合酶途径被同化成合氨酸。

formation of one molecule of H_2. In free-living diazotrophs, the nitrogenase-generated(固氮酶生成) ammonium is assimilated into glutamate through the glutamine synthetase(谷氨酰胺合成酶)/glutamate synthase(谷氨酸合酶) pathway.

固氮酶活力对氧气敏感。

Enzymes responsible for nitrogenase action are very susceptible(易感) to destruction by oxygen. (In fact, many bacteria cease production of the enzyme in the presence of oxygen). Many nitrogen-fixing organisms(有机体) exist only in anaerobic conditions(厌氧条件), respiring to draw down oxygen levels, or binding the oxygen with a protein such as leghemoglobin(豆血红蛋白).

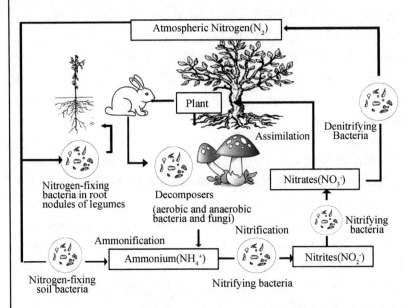

Schematic representation of the nitrogen cycle.
Abiotic nitrogen fixation has been omitted.

4.2 Amino acid synthesis 氨基酸合成

氨基酸合成是一系列生化过程(代谢途径),各种氨基酸由其他化合物合成。

Amino acid synthesis is the set of biochemical processes (metabolic pathways) by which the various amino acids are produced from other compounds. The substrates(底物) for these processes are various compounds in the organism's(有机体) diet or growth media(生长介质). Not all organisms are able to synthesize all amino acids. For example, humans are able to synthesize only 12 of the 20 standard amino acids.

A fundamental(基本的) problem for biological systems is to obtain nitrogen in an easily usable(可用的) form. This problem is solved by certain microorganisms capable of reducing the inert N≡N molecule (nitrogen gas) to two molecules of ammonia in one of the most remarkable reactions in biochemistry. Ammonia is the source of nitrogen for all the amino acids. The carbon backbones(碳骨架) come from the glycolytic(糖酵解) pathway, the pentose phosphate(戊糖磷酸途径) pathway, or the citric acid cycle(三羧酸循环).

In amino acid production, one encounters(遇到) an important problem in biosynthesis, namely stereochemical(立体化学的) control. Because all amino acids except glycine(甘氨酸) are chiral(手性), biosynthetic(生物合成) pathways must generate the correct isomer(同分异构体) with high fidelity(保真度). In each of the 19 pathways for the generation of chiral amino acids, the stereochemistry at the α-carbon atom is established by a transamination(转氨作用) reaction that involves pyridoxal(吡哆醛) phosphate. Almost all the transaminases that catalyze these reactions descend from a common ancestor(祖先), illustrating(说明) once again that effective solutions to biochemical problems are retained throughout evolution.

Biosynthetic pathways are often highly regulated such that building-blocks(结构单元) are synthesized only when supplies are low. Very often, a high concentration(浓度) of the final product of a pathway inhibits(抑制) the activity of enzymes that function early in the pathway. Often present are allosteric enzymes capable of sensing and responding to concentrations of regulatory species. These enzymes are similar in functional properties to aspartate transcarbamoylase(天冬氨酸) and its regulators. Feedback(反馈) and allosteric mechanisms(变构机

制) ensure that all twenty amino acids are maintained in sufficient (足够) amounts for protein synthesis and other processes.

Amino acid synthesis 氨基酸合成

Amino acids are synthesized from α-ketoacids, and later transaminated from another amino acid, usually Glutamate(谷氨酸). The enzyme involved in this reaction is an aminotransferase(转氨酶).

$$\alpha\text{-ketoacid} + \text{glutamate} \rightleftharpoons \text{amino acid} + \alpha\text{-ketoglutarate}$$

Glutamate itself is formed by amination of α-ketoglutarate:

$$\alpha\text{-ketoglutarate} + NH \rightleftharpoons \text{glutamate}$$

Nitrogen fixation: Microorganisms use ATP and a powerful reductant to reduce atmospheric nitrogen to ammonia

Microorganisms use ATP and reduced ferredoxin(铁氧还蛋白), a powerful reductant, to reduce N_2 to NH_3. An iron-molybdenum(铁钼) cluster in nitrogenase deftly catalyzes the fixation of N_2, a very inert molecule. Higher organisms consume the fixed nitrogen to synthesize amino acids, nucleotides, and other nitrogen-containing biomolecules. The major points of entry of NH_4^+ into metabolism are glutamine or glutamate(谷氨酰胺和谷氨酸).

Amino acids are made from intermediates of the citric acid cycle and other major pathways

Of the basic set of 20 amino acids (not counting selenocysteine (硒代半胱氨酸)), there are 8 that human beings cannot synthesize. In addition, the amino acids arginine(精氨酸), cysteine(半胱氨酸), glycine(甘氨酸), glutamine(谷氨酰胺), histidine(组氨酸), proline(脯氨酸), serine(丝氨酸), and tyrosine(酪氨酸) are considered conditionally essential, meaning they are not normally required in the diet, but must be supplied exogenously(外部) to specific populations that do not synthesize it in adequate amounts.

For example, enough arginine is synthesized by the urea

cycle to meet the needs of an adult but perhaps not those of a growing child. Amino acids that must be obtained from the diet are called essential amino acids. Nonessential amino acids are produced in the body. The pathways for the synthesis of nonessential amino acids are quite simple. Glutamate dehydrogenase（谷氨酸脱氢酶）catalyzes the reductive amination of α-ketoglutarate to glutamate. A transamination reaction takes place in the synthesis of most amino acids. At this step, the chirality of the amino acid is established. Alanine and aspartate are synthesized by the transamination of pyruvate（丙酮酸）and oxaloacetate（草酰乙酸）, respectively. The pathways for the biosynthesis of essential amino acids are much more complex than those for the nonessential ones. Glutamine is synthesized from NH_4^+ and glutamate, and asparagine（天冬酰胺）is synthesized similarly. Proline and arginine are derived from glutamate. Serine, formed from 3-phosphoglycerate, is the precursor（前体）of glycine and cysteine. Tyrosine is synthesized by the hydroxylation of phenylalanine（羟化苯丙氨酸）, an essential amino acid.

Tetrahydrofolate（四氢叶酸）, a carrier of activated one-carbon units, plays an important role in the metabolism of amino acids and nucleotides. This coenzyme（辅酶）carries one-carbon units at three oxidation states, which are interconvertible（可以相互转化）: most reduced—methyl（甲基）; intermediate（中级）—methylene（甲烷）; and most oxidized—formyl（甲酰基）, formimino, and methenyl（次甲基）. The major donor（供体）of activated methyl groups is S-adenosylmethionine（S-腺苷蛋氨酸）, which is synthesized by the transfer of an adenosyl（腺苷）group from ATP to the sulfur atom of methionine. S-Adenosylhomocysteine（S-腺苷同型半胱氨酸）is formed when the activated methyl group is transferred to an acceptor. It is hydrolyzed to adenosine and homocysteine（同型半胱氨酸）, the latter（后者）of which is then methylated to methionine（甲硫氨酸）to complete the activated methyl cycle. Cortisol（皮质醇）inhibits protein synthesis.

四氢叶酸，被激活的一碳单位的载体，在氨基酸和核苷酸代谢中有重要作用。

Amino acid biosynthesis is regulated by feedback inhibition

Most of the pathways of amino acid biosynthesis are regulated by feedback inhibition, in which the committed step is allosterically inhibited by the final product. Branched pathways require extensive interaction(广泛的相互作用) among the branches that includes both negative and positive regulation. The regulation of glutamine synthetase(谷氨酰胺合成酶) from *E. coli* is a striking(引人注目) demonstration(示范) of cumulative feedback inhibition and of control by a cascade of reversible(可逆) covalent modifications.

Amino acids are precursors of many biomolecules

Amino acids are precursors of a variety of biomolecules. Glutathione (γ-Glu-Cys-Gly) serves as a sulfhydryl buffer and detoxifying agent(解毒剂). Glutathione peroxidase, a selenoenzyme(含硒酶), catalyzes the reduction of hydrogen peroxide and organic peroxides by glutathione. Nitric oxide(一氧化氮), a short-lived messenger, is formed from arginine. Porphyrins(嘌呤类化合物) are synthesized from glycine and succinyl CoA, which condense to give δ-aminolevulinate. Two molecules of this intermediate become linked to form porphobilinogen. Four molecules of porphobilinogen combine to form a linear(线性) tetrapyrrole(四吡咯), which cyclizes(使环化) to uroporphyrinogen(尿卟啉原) III. Oxidation and side-chain modifications lead to the synthesis of protoporphyrin(原卟啉) IX, which acquires an iron atom to form heme(血红素).

4.3 Nucleotides synthesis 核苷酸合成

Nucleotides are molecules that, when joined together, make up the structural units of RNA and DNA. In addition, nucleotides play central roles in metabolism, in which capacity they serve as sources of chemical energy (adenosine triphosphate 三磷酸腺苷 and guanosine triphosphate 鸟嘌呤三磷酸核苷), participate in cellular signaling (cyclic guanosine monophosphate 环磷酸鸟苷 and cyclic adenosine monophosphate 环磷酸腺苷), and are incorporated into important cofactors of enzymatic reactions (coenzyme A, flavin adenine dinucleotide（黄素腺嘌呤二核苷酸）, flavin mononucleotide（黄素单核苷酸）, and nicotinamide adenine dinucleotide phosphate 烟酰胺腺嘌呤二核苷酸磷酸).

4.3.1　Nucleotide structure 核苷酸结构

A nucleotide is composed of a nucleobase（碱基）(nitrogenous base), a five-carbon sugar (either ribose or 2′-deoxyribose), and one to three phosphate groups. Together, the nucleobase and sugar comprise a nucleoside. The phosphate groups form bonds with either the 2, 3, or 5-carbon of the sugar, with the 5-carbon site most common. Cyclic nucleotides form when the phosphate group is bound to two of the sugar's hydroxyl（羟基）groups. Ribonucleotides（核糖核酸）are nucleotides where the sugar is ribose, and deoxyribonucleotides（脱氧核糖核酸）contain the sugar deoxyribose. Nucleotides can contain either a purine（嘌呤）or a pyrimidine（嘧啶）base. Nucleic acids are polymeric macromolecules made from nucleotide monomers. In DNA, the purine bases are adenine and guanine（腺嘌呤和鸟嘌呤）, while the pyrimidines are thymine and cytosine（胸腺嘧啶和胞嘧啶）. RNA uses uracil（尿嘧啶）in place of thymine. Adenine always pairs with thymine by 2 hydrogen bonds, while guanine pairs with cytosine through 3 hydrogen bonds, each due to their unique structures.

一个核苷酸由一个碱基（含氮碱基），一个五碳糖（核糖或 2′-脱氧核糖）和一个到三个磷酸基团组成。

核酸是由核苷酸单体形成的聚合物。

DNA 和 RNA 结构上的区别。

Structural elements of the most common nucleotides.

4.3.2 Synthesis 合成

Nucleotides can be synthesized by a variety of means both in vitro(体外) and in vivo(体内). In vivo, nucleotides can be synthesized de novo or recycled（回收） through salvage pathways. Nucleotides undergo breakdown such that useful parts can be reused in synthesis reactions to create new nucleotides. In vitro, protecting groups may be used during laboratory production of nucleotides. A purified（纯化） nucleoside is protected to create a phosphoramidite, which can then be used to obtain analogues(类似物) not found in nature and/or to synthesize an oligonucleotide(寡核苷酸).

在体内和体外可以通过各种手段合成核苷酸。

The synthesis of Uridine monophosphate UMP.

嘧啶核苷酸合成。

(1) Pyrimidine ribonucleotides 嘧啶核糖核苷酸

Pyrimidine nucleotide synthesis starts with the formation of carbamoyl phosphate from glutamine(谷氨酰胺) and CO_2.

The cyclization reaction between carbamoyl phosphate reacts with aspartate, yielding orotate(乳清酸) in subsequent(随后) steps. Orotate reacts with 5-phosphoribosyl α-diphosphate (PRPP), yielding orotidine monophosphate(乳清核苷酸) (OMP), which is decarboxylated to form uridine monophosphate (UMP). It is from UMP that other pyrimidine nucleotides are derived. UMP is phosphorylated to uridine triphosphate (UTP) via two sequential reactions with ATP. Cytidine monophosphate (CMP) is derived from conversion of UTP to cytidine triphosphate (CTP) with subsequent loss of two phosphates.

(2) Purine ribonucleotides 嘌呤核糖核苷酸

The atoms which are used to build the purine nucleotides come from a variety of sources:

嘌呤环的生物合成,及环上原子的来源。

	The biosynthetic origins of purine ring atoms
	N_1 arises from the amine group of Asp C_2 and C_8 originate from formate N_3 and N_9 are contributed by the amide group of Gln C_4, C_5 and N_7 are derived from Gly C_6 comes from HCO_3^- (CO_2)

The de novo synthesis of purine nucleotides by which these precursors are incorporated(合并) into the purine ring proceeds by a 10-step pathway to the branch-point intermediate(中间) IMP, the nucleotide of the base hypoxanthine. AMP and GMP are subsequently synthesized from this intermediate via separate, two-step pathways. Thus, purine moieties(部分) are initially formed as part of the ribonucleotides rather than as free bases.

Six enzymes take part in IMP synthesis. Three of them are multifunctional:

六种酶参与 IMP 合成,其中三种是多功能的。

- GART (reactions 2, 3, and 5)
- PAICS (reactions 6, and 7)

● ATIC (reactions 9, and 10)

The synthesis of IMP.

IMP 的生物合成。

The pathway starts with the formation of PRPP. PRPS1 is the enzyme that activates R5P, which is formed primarily by the pentose phosphate pathway, to PRPP by reacting it with ATP. The reaction is unusual in that a pyrophosphoryl group is directly transferred from ATP to C_1 of R5P and that the product has the α configuration about C_1. This reaction is also shared with the pathways for the synthesis of Trp, His, and the pyrimidine nucleotides. Being on a major metabolic crossroad and requiring much energy, this reaction is highly regulated.

合成途径从 PRPP 的形成开始。PRPS1 酶激活 R5P,其主要由磷酸戊糖途径生成。

In the first reaction unique to purine nucleotide biosynthesis, PPAT catalyzes the displacement(位移) of PRPP's pyrophosphate group (PP_i) by Gln's amide nitrogen(谷氨酰胺的酰胺态氮). The reaction occurs with the inversion of configuration(结构) about

嘌呤核苷酸生物合成第一步反应中,PPAT 催化 PRPP 焦磷酸基团的位移。

ribose C_1, thereby forming β-5-phosphorybosylamine (5-PRA) and establishing the anomeric(异头碳) form of the future nucleotide. This reaction, which is driven to completion by the subsequent hydrolysis of the released PP_i, is the pathway's flux-generating step and is therefore regulated, too.

(3) Length unit 长度单位

Nucleotide (abbreviated 缩写"nt") is a common length unit for single-stranded RNA, similar to how base pair is a length unit for double-stranded DNA.

(4) Abbreviation codes for degenerate bases 简并碱基的缩写

The IUPAC has designated the symbols for nucleotides. Apart from the five (A, G, C, T/U) bases, often degenerate bases are used especially for designing PCR primers. These nucleotide codes are listed here.

IUPAC nucleotide code	Base
A	Adenine
C	Cytosine
G	Guanine
T (or U)	Thymine (or Uracil)
R	A or G
Y	C or T (U)
S	G or C
W	A or T (U)
K	G or T (U)
M	A or C
B	C or G or T (U)
D	A or G or T (U)
H	A or C or T (U)
V	A or C or G
N	any base
. or -	gap

4.4 Urea cycle 尿素循环

The urea cycle(尿素循环) (also known as the ornithine 鸟氨酸 cycle) is a cycle of biochemical reactions occurring in many animals that produces urea ($(NH_2)_2CO$) from ammonia (NH_3). This cycle was the first metabolic cycle discovered (Hans Krebs and Kurt Henseleit, 1932), 5 years before the discovery of the TCA cycle. In mammals(哺乳动物), the urea cycle takes place primarily in the liver, and to a lesser extent in the kidney.

4.4.1 Function 功能

Organisms that cannot easily and quickly remove ammonia usually have to convert it to some other substance, like urea or uric acid, which are much less toxic(有毒的). Insufficiency of the urea cycle occurs in some genetic disorders (inborn errors of metabolism), and in liver failure. The result of liver failure is accumulation of nitrogenous waste, mainly ammonia, which leads to hepatic encephalopathy(肝性脑病).

4.4.2 Reactions 反应

The urea cycle consists of five reactions: two mitochondrial(线粒体) and three cytosolic(胞浆). The cycle converts two amino groups, one from NH_4^+ and one from Asp, and a carbon atom from HCO_3^-, to the relatively nontoxic(没有毒性的) excretion(排泄) product urea at the cost of four "high-energy" phosphate bonds (3 ATP hydrolyzed to 2 ADP and one AMP). Ornithine(鸟氨酸) is the carrier of these carbon and nitrogen atoms.

Reactions of the urea cycle.

Step	Reactants	Products	Catalyzed by	Location
1	$NH_4^+ + HCO_3^-$ + 2ATP	carbamoyl phosphate + 2ADP + P_i	CPS1	mitochondria

(续表)

Step	Reactants	Products	Catalyzed by	Location
2	carbamoyl phosphate + ornithine	citrulline + P_i	OTC	mitochondria
3	citrulline + aspartate + ATP	argininosuccinate + AMP + PP_i	ASS	cytosol
4	argininosuccinate	Arg + fumarate	ASL	cytosol
5	Arg + H_2O	ornithine + urea	ARG1	cytosol

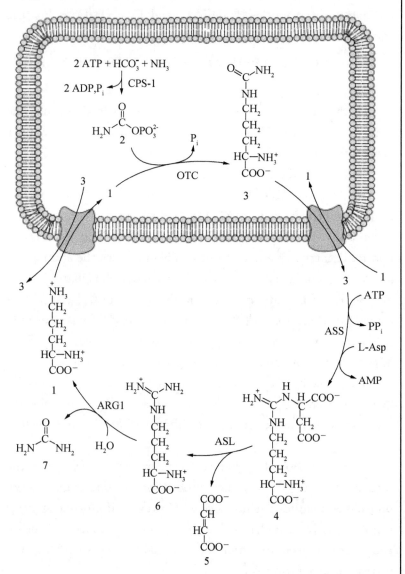

The reactions of the urea cycle.

尿素循环反应示意图。
1 L-ornithine
2 carbamoyl phosphate
3 L-citrulline
4 argininosuccinate
5 fumarate(富马酸)
6 L-arginine
7 urea
L-Asp L-aspartate
CPS-1 carbamoyl phosphate synthetase I
OTC Ornithine transcarbamylase
ASS argininosuccinate synthetase
ASL argininosuccinate lyase
ARG1 arginase 1

In the first reaction, $NH_4^+ + HCO_3^-$ is equivalent to $NH_3 + CO_2 + H_2O$.

Thus, the overall equation of the urea cycle is:

$$NH_3 + CO_2 + aspartate + 3ATP + 2H_2O \longrightarrow urea + fumarate + 2ADP + 2P_i + AMP + PP_i$$

Since fumarate is obtained by removing NH_3 from aspartate (by means of reactions 3 and 4), and $PP_i + H_2O \rightarrow 2P_i$, the equation can be simplified as follows:

$$2NH_3 + CO_2 + 3ATP + H_2O \longrightarrow urea + 2ADP + 4P_i + AMP$$

产生 NADH 的两种方式。

Note that reactions related to the urea cycle also cause the reduction of 2 NADH, so the urea cycle releases slightly more energy than it consumes.

These NADH are produced in two ways:

一个 NADH 分子减少是由于谷氨酸脱氢酶作用下,谷氨酸转化为铵态氮和 α-酮戊二酸。

● One NADH molecule is reduced by the enzyme glutamate dehydrogenase(谷氨酸脱氢酶) in the conversion of glutamate to ammonium and α-ketoglutarate. Glutamate is the non-toxic carrier of amine groups. This provides the ammonium ion used in the initial synthesis of carbamoyl phosphate.

分布在胞浆内的延胡索酸在胞浆延胡索酸作用下转化为苹果酸。这个苹果酸再转为草酰乙酸,在胞质中的 NADH 浓度下降。

● The fumarate released in the cytosol(细胞溶液) is converted to malate(苹果酸) by cytosolic(胞浆) fumarase(延胡索酸酶). This malate is then converted to oxaloacetate(草酰乙酸盐) by cytosolic malate dehydrogenase, generating a reduced NADH in the cytosol. Oxaloacetate is one of the keto acids(酮酸) preferred by transaminases, and so will be recycled to aspartate, maintained the flow of nitrogen into the urea cycle.

产生两个 NADH 能生成 5 个 ATP,并且为尿素循环提供高能磷酸键。

The two NADH produced can provide energy for the formation of 5 ATP, a net production of one high energy phosphate bond for the urea cycle. However, if gluconeogenesis(糖异生作用) is underway in the cytosol, the latter reducing equivalent is used to drive the reversal of the GAPDH step instead of generating ATP.

草酰乙酸的去向是通过转氨作用生成天冬氨酸,或者转化成磷酸烯醇式丙酮酸。

The fate of oxaloacetate is either to produce aspartate via transamination or to be converted to phosphoenol pyruvate, which is a substrate to glucose. An excellent way to memorize the urea cycle is to remember the mnemonic phrase(记忆短语) "Ordinarily Careless Crappers Are Also Frivolous About Urination." The first letter of each word corresponds to the first letter of each of the main reactants or products that are combined with each other or produced as one progresses through the five reactions of the cycle (Ornithine, Carbamoyl phosphate, Citrulline, Aspartate, Argininosuccinate, Fumarate, Arginine, Urea, 鸟氨酸, 氨基甲酰磷酸, 瓜氨酸, 天门冬氨酸, 精氨酸, 富马酸, 精氨酸, 尿素).

4.4.3 Regulation 调节

(1) N-Acetylglutamic acid N-乙酰谷氨酸

The synthesis of carbamoyl phosphate and the urea cycle are dependent on the presence of NAcGlu, which allosterically activates CPS1. Synthesis of NAcGlu by NAGS, is stimulated by Arg, allosteric stimulator of NAGS, and Glu, a product in the transamination reactions and one of NAGS's substrates, both of which are elevated when free amino acids are elevated. So Glu is not only a substrate for the urea cycle reactions but also serves as an activator for the urea cycle.

氨甲酰磷酸的合成和尿素循环是依赖于NAcGlu,其通过别构效应激活CPS1。

(2) Substrate concentrations 底物浓度

The remaining enzymes of the cycle are controlled by the concentrations of their substrates. Thus, inherited deficiencies in the cycle enzymes other than ARG1 do not result in significant decrease in urea production (the total lack of any cycle enzyme results in death shortly after birth). Rather, the deficient enzyme's substrate builds up, increasing the rate of the deficient reaction to normal.

在循环中,酶活性受其底物浓度的影响。

The anomalous(异常) substrate buildup is not without cost, however. The substrate concentrations become elevated all the way back up the cycle to NH_4^+, resulting in hyperammonemia (elevated $[NH_4^+]_P$). Although the root cause of NH_4^+ toxicity is not completely understood, a high $[NH_4^+]$

尽管 NH_4^+ 毒性的根本原因尚不完全清楚,但高 $[NH_4^+]$ 会对大脑有影响,如发育迟滞和嗜睡。

puts an enormous strain on the NH_4^+-clearing system, especially in the brain (symptoms of urea cycle enzyme deficiencies include mental retardation 智力迟钝 and lethargy 昏睡). This clearing system involves GLUD1 and GLUL, which decrease the 2OG and Glu pools. The brain is most sensitive to the depletion of these pools. Depletion of 2OG decreases the rate of TCAC, whereas Glu is both a neurotransmitter(神经递质) and a precursor to GABA, another neurotransmitter.

【Key words】

nitrogen fixation 固氮作用	ornithine 鸟氨酸
orotate 乳清酸	uridine monophosphate (UMP) 尿苷一磷酸
degenerate bases 简并碱基	IMP 次黄嘌呤核苷酸
adenine 腺嘌呤	carbamoyl phosphate 氨基甲酰磷酸
guanine 鸟嘌呤	citrulline 瓜氨酸
cytosine 胞嘧啶	aspartate 天冬氨酸
Thymine 胸腺嘧啶	argininosuccinate 精氨
uracil 尿嘧啶	fumarate 富马酸
urea cycle 尿素循环	arginine 精氨酸

Questions

1. Define nitrogen fixation. What organisms are capable of nitrogen fixation?
2. Name the α-ketoacid that is formed by the transamination of each of the following amino acids:
   (a) Alanine
   (b) Aspartate
   (c) Glutamate
   (d) Leucine
   (e) Phenylalanine
   (f) Tyrosine
3. Identify the source of the atoms in the pyrimidine ring.
4. Write a balanced equation for the synthesis of orotate from glutamine, CO_2, and aspartate.

References

[1] McKee Trudy, R McKee James. Biochemistry: An Introduction (Second Edition)[M]. New York: McGraw-Hill Companies, 1999.

[2] Nelson D L, Cox M M. Lehninger Principles of Biochemistry (Third Edition)[M]. Derbyshire: Worth Publishers, 2000.

[3] Ding W, Jia H T. Biochemistry (First Edition)[M]. Beijing: Higher Education Press, 2012.

[4] 郑集，陈钧辉. 普通生物化学(第四版)[M]. 北京：高等教育出版社，2007.

[5] Berg J M, Tymoczko J L, Styer L. Biochemistry (Seventh Edition)[M]. New York: W. H. Freeman and Company, 2012.

Chapter 5　Oxidative phosphorylation 氧化磷酸化

5.1　Overview of oxidative phosphorylation 氧化磷酸化概述

氧化磷酸化的定义。
氧化磷酸化产生 ATP 并释放能量。

Oxidative phosphorylation is a metabolic pathway that uses energy released by the oxidation of nutrients(营养素) to produce adenosine triphosphate（三磷酸腺苷，ATP）. Although the many forms of life on earth use a range of different nutrients, almost all aerobic organisms carry out oxidative phosphorylation to produce ATP, the molecule that supplies energy to metabolism. This pathway is probably so pervasive because it is a highly efficient way of releasing energy, compared to alternative fermentation processes(发酵过程) such as anaerobic glycolysis(无氧糖酵解).

电子传递链示意图。

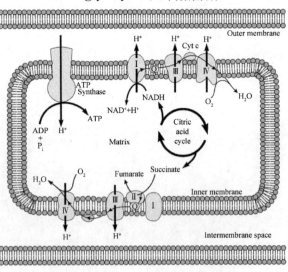

The electron transport chain in the mitochondrion is the site of oxidative phosphorylation in eukaryotes. The NADH and succinate generated in the citric acid cycle are oxidized, releasing energy to power the ATP synthase.

During oxidative phosphorylation, electrons are transferred from electron donors(供体) to electron acceptors(受体) such as oxygen, in redox reactions(氧化还原反应). These redox reactions release energy, which is used to form ATP. In eukaryotes, these redox reactions are carried out by a series of protein complexes within mitochondria, whereas, in prokaryotes, these proteins are located in the cell's inner membranes. These linked sets of proteins are called electron transport chains(电子传递链). In eukaryotes, five main protein complexes are involved, whereas in prokaryotes many different enzymes are present, using a variety of electron donors and acceptors.

The energy released by electrons flowing through this electron transport chain is used to transport protons(质子) across the inner mitochondrial membrane, in a process called chemiosmosis(化学渗透). This generates potential energy in the form of a pH gradient(梯度) and an electrical potential across this membrane. This store of energy is tapped by allowing protons to flow back across the membrane and down this gradient, through a large enzyme called ATP synthase(合酶). This enzyme uses this energy to generate ATP from adenosine diphosphate(二磷酸腺苷, ADP), in a phosphorylation reaction. This reaction is driven by the proton flow, which forces the rotation(循环) of a part of the enzyme; the ATP synthase is a rotary mechanical motor(旋转的机械马达).

Although oxidative phosphorylation is a vital part of metabolism, it produces reactive oxygen species(活性氧) such as superoxide(过氧化物) and hydrogen peroxide(过氧化氢), which lead to propagation(产生) of free radicals(自由基), damaging cells and contributing to disease and, possibly, aging (老化, senescence). The enzymes carrying out this metabolic pathway are also the target of many drugs and poisons that inhibit their activities.

5.2 Overview of energy transfer by chemiosmosis
通过化学渗透能量转移概述

Oxidative phosphorylation(氧化磷酸化) works by using energy-releasing chemical reactions(放能化学反应) to drive energy-requiring reactions(需能化学反应): The two sets of reactions are said to be coupled. This means one cannot occur without the other. The flow of electrons through the electron transport chain, from electron donors such as NADH to electron acceptors such as oxygen, is an exergonic process(放能过程)—it releases energy, whereas the synthesis of ATP is an endergonic process(吸能过程), which requires an input of energy. Both the electron transport chain and the ATP synthase are embedded in a membrane, and energy is transferred from electron transport chain to the ATP synthase by movements of protons across this membrane, in a process called *chemiosmosis*. In practice, this is like a simple electric circuit(电路), with a current of protons being driven from the negative N-side of the membrane to the positive P-side by the proton—pumping enzymes(质子-泵酶) of the electron transport chain. These enzymes are like a battery, as they perform work to drive current through the circuit. The movement of protons creates an electrochemical gradient(电化学梯度) across the membrane, which is often called the *proton-motive force*(质子动力). This gradient has two components: a difference in proton concentration (a H^+ gradient) and a difference in electric potential(电势), with the N-side having a negative charge. The energy is stored largely as the difference of electric potentials in mitochondria, but also as a pH gradient in chloroplasts(叶绿体).

ATP synthase releases this stored energy by completing the circuit and allowing protons to flow down the electrochemical gradient, back to the N-side of the membrane. This enzyme is like an electric motor(电动机) as it uses the proton-motive force

（质子动力势）to drive（驱动）the rotation（旋转）of part of its structure and couples（偶联）this motion to the synthesis of ATP.

The amount of energy released by oxidative phosphorylation is high, compared with the amount produced by anaerobic fermentation. Glycolysis produces only 2 ATP molecules, but somewhere between 30 and 36 ATPs are produced by the oxidative phosphorylation of the 10 NADH and 2 succinate molecules made by converting one molecule of glucose to carbon dioxide and water, while each cycle of beta oxidation of a fatty acid yields about 14 ATPs. These ATP yields are theoretical maximum values（理论最大值）; in practice, some protons leak across the membrane, lowering the yield of ATP.

氧化磷酸化释放的能量很高。

糖酵解只产生2个ATP，而氧化磷酸化可产生30~36个ATP。

5.3 Electron and proton transfer molecules and electron transport chains 电子和质子转移分子和电子传递链

5.3.1 Electron and proton transfer molecules 电子和质子转移分子

Reduction of coenzyme Q from its ubiquinone form (Q) to the reduced ubiquinol form (QH_2).

辅酶Q的泛醌结构及其对电子的传递。

| 细胞色素 C 在线粒体电子转移中的作用。

The electron transport chain carries both protons and electrons, passing electrons from donors to acceptors, and transporting protons across a membrane. These processes use both soluble and protein-bound transfer molecules(蛋白结合转移分子). In mitochondria, electrons are transferred within the intermembrane space(膜间隙) by the water-soluble electron transfer protein cytochrome C(细胞色素 C). This carries only electrons, and these are transferred by the reduction and oxidation of an iron atom(铁原子) that the protein holds within a heme group(血红素) in its structure. Cytochrome C is also found in some bacteria, where it is located within the periplasmic space(周质空间).

| 氧化还原循环中辅酶 Q_{10} 的作用。

Within the inner mitochondrial membrane, the lipid-soluble electron carrier coenzyme Q_{10} (Q) carries both electrons and protons by a redox cycle(氧化还原循环). This small benzoquinone molecule(苯醌分子) is very hydrophobic(疏水的), so it diffuses freely within the membrane. When Q accepts two electrons and two protons, it becomes reduced to the ubiquinol(泛醌) form (QH_2); when QH_2 releases two electrons and two protons, it becomes oxidized back to the ubiquinone(辅酶)(Q) form. As a result, if two enzymes are arranged so that Q is reduced on one side of the membrane and QH_2 oxidized on the other, ubiquinone will couple these reactions and shuttle protons across the membrane. Some bacterial electron transport chains use different quinones(醌类), such as menaquinone(甲基萘醌类), in addition to ubiquinone.

| Q 和 QH_2 之间的相互转换。

| 黄素、铁硫簇合物和细胞色素等其他电子传递物。

Within proteins, electrons are transferred between flavin cofactors(黄素辅助因子), iron-sulfur clusters(铁硫簇合物), and cytochromes. There are several types of iron-sulfur cluster. The simplest kind found in the electron transfer chain consists of two iron atoms joined by two atoms of inorganic sulfur(无机硫); these are called [2Fe-2S] clusters. The second kind, called [4Fe-4S], contains a cube of four iron atoms and four sulfur atoms. Each iron atom in these clusters is coordinated by an additional amino acid, usually by the sulfur atom of cysteine(半胱氨酸). Metal ion cofactors undergo redox reactions without binding or releasing protons, so in the electron transport chain they serve solely to transport electrons through proteins. Electrons move quite long distances through proteins by hopping along chains(跳跃链) of these cofactors. This occurs by quantum tunnelling(量子穿隧效应), which is rapid over distances of less than 1.4×10^{-9} m.

5.3.2 Eukaryotic electron transport chains 真核生物的电子传递链

Many catabolic biochemical processes, such as glycolysis(糖酵解), the citric acid cycle(三羧酸循环), and β oxidation(β氧化), produce the reduced coenzyme(还原型辅酶) NADH. This coenzyme contains electrons(电子) that have a high transfer potential; in other words, they will release a large amount of energy upon oxidation(氧化). However, the cell does not release this energy all at once, as this would be an uncontrollable reaction. Instead, the electrons are removed from NADH and passed to oxygen through a series of enzymes that each release a small amount of the energy. This set of enzymes(一系列酶), consisting of complexes I through IV, is called the electron transport chain(电子传递链) and is found in the inner membrane of the mitochondrion(线粒体). Succinate(琥珀酸) is also oxidized by the electron transport chain, but feeds into the pathway at a different point.

电子传递链位于线粒体内膜上。

In eukaryotes, the enzymes in this electron transport system use the energy released from the oxidation of NADH to pump(泵出) protons across the inner membrane of the mitochondrion. This causes protons to build up in the intermembrane space, and generates an electrochemical gradient across the membrane. The energy stored in this potential is then used by ATP synthase to produce ATP. Oxidative phosphorylation in the eukaryotic mitochondrion is the best-understood example of this process. The mitochondrion is present in almost all eukaryotes, with the exception of anaerobic protozoa(厌氧原生动物) such as *Trichomonas vaginalis*(阴道毛滴虫) that instead reduce protons to hydrogen

在真核生物中,电子传递链中的酶,利用 NADH 氧化释放的能量,将质子从线粒体内膜内侧泵出,形成跨膜电化学梯度。

in a remnant mitochondrion（残余线粒体）called a hydrogenosome（氢化酶体）.

Typical respiratory enzymes and substrates in eukaryotes

Respiratory enzyme	Redox pair（氧化还原对）	Midpoint potential (Volts)
NADH dehydrogenase	NAD^+/NADH	-0.32
Succinate dehydrogenase	FMN or FAD/$FMNH_2$ or $FADH_2$	-0.20
Cytochrome bc_1 complex	Coenzyme $Q10_{ox}$/Coenzyme $Q10_{red}$	$+0.06$
Cytochrome complex	Cytochrome b_{ox}/Cytochrome b_{red}	$+0.12$
Complex IV	Cytochrome c_{ox}/Cytochrome c_{red}	$+0.22$
Complex IV	Cytochrome a_{ox}/Cytochrome a_{red}	$+0.29$
Complex IV	O_2/HO^-	$+0.82$
Conditions：pH=7		

NADH dehydrogenase：NADH 脱氢酶
Succinate dehydrogenase：琥珀酸脱氢酶
Cytochrome bc_1 complex：细胞色素 bc_1 复合体

（1）NADH‐coenzyme Q oxidoreductase（complex I） NADH‐辅酶 Q 氧化还原酶（复合体 I）

NADH‐coenzyme Q oxidoreductase, also known as *NADH dehydrogenase* or *complex I*, is the first protein in the electron transport chain. Complex I is a giant enzyme with the mammalian（哺乳类）complex I having 46 subunits（亚基）and a molecular mass of about 1,000 kilodaltons (kDa). The structure is known in detail only from a bacterium; in most organisms the complex resembles a boot with a large "ball" poking out from the membrane into the mitochondrion. The genes that encode the individual proteins are contained in both the cell nucleus（细胞核）and the mitochondrial genome（基因组）, as is the case for many enzymes present in the mitochondrion.

复合体 I 是电子传递链中的第一个蛋白质酶。

The reaction that is catalyzed by this enzyme is the two electron reduction by NADH of coenzyme Q_{10} or ubiquinone (represented as Q in the equation below), a lipid-soluble quinone that is found in the mitochondrion membrane:

复合体 I 催化的反应。

$$NADH + Q + 5H^+_{matrix} \longrightarrow NAD^+ + QH_2 + 4H^+_{intermembrane}$$

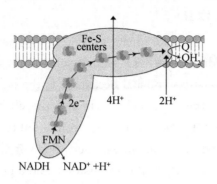

Complex I or NADH‑Q oxidoreductase. In all diagrams of respiratory complexes in this chapter, the matrix is at the bottom, with the intermembrane space above.

The start of the reaction, and indeed of the entire electron chain, is the binding of a NADH molecule to complex I and the donation of two electrons. The electrons enter complex I via a prosthetic group（辅基） attached to the complex, flavin mononucleotide（黄素单核苷酸，FMN）. The addition of electrons to FMN converts it to its reduced form（还原型）, FMNH2. The electrons are then transferred through a series of iron-sulfur clusters（铁-硫簇）: the second kind of prosthetic group（辅基）present in the complex. There are both [2Fe‑2S] and [4Fe‑4S] iron-sulfur clusters in complex I.

电子通过一系列的铁硫簇合物转移。

As the electrons pass through this complex, four protons are pumped from the matrix into the intermembrane space. Exactly how this occurs is unclear, but it seems to involve conformational（构象的）changes in complex I that cause the protein to bind protons on the N‑side of the membrane and release them on the P-side of the membrane. Finally, the electrons are transferred from the chain of iron-sulfur clusters to a ubiquinone molecule（泛醌分子）in the membrane. Reduction of ubiquinone also contributes to the generation of a proton gradient, as two protons are taken up from the matrix as it is reduced to ubiquinol（QH_2）.

当电子通过这个复合物，4个质子泵出线粒体内侧，进入膜间隙，该过程涉及复合体I构象的改变。

(2) Succinate‑Q oxidoreductase (complex II) 琥珀酸‑Q氧化还原酶(复合体 II)

复合体 II 是电子传递链的第二个入口点,也是唯一一个既属于三羧酸循环又属于电子传递链的酶。

Succinate‑Q oxidoreductase, also known as complex II or succinate dehydrogenase, is a second entry point to the electron transport chain. It is unusual because it is the only enzyme that is part of both the citric acid cycle and the electron transport chain. Complex II consists of four protein subunits(亚基) and contains a bound flavin adenine dinucleotide(黄素腺嘌呤二核苷酸,FAD) cofactor, iron-sulfur clusters, and a heme group that does not participate in electron transfer to coenzyme Q, but is believed to be important in decreasing production of reactive oxygen species. It oxidizes succinate to fumarate(延胡索酸) and reduces ubiquinone. As this reaction releases less energy than the oxidation of NADH, complex II does not transport protons across the membrane and does not contribute to the proton gradient.

$$Succinate + Q \longrightarrow Fumarate + QH_2$$

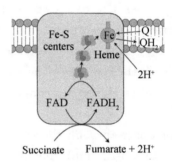

Complex II: Succinate‑Q oxidoreductase.

某些真核生物,例如寄生虫蛔虫,含有延胡索酸还原酶,具有类似复合体 II 的功能。

In some eukaryotes(真核生物), such as the parasitic worm *Ascaris suum*(寄生虫蛔虫), an enzyme similar to complex II, fumarate reductase(延胡索酸还原酶) (menaquinol: fumarate oxidoreductase, or QFR), operates in reverse to oxidize ubiquinol and reduce fumarate. This allows the worm to survive in the anaerobic environment of the large intestine(大肠), carrying out anaerobic oxidative phosphorylation with fumarate as the electron acceptor. Another unconventional function of complex II is seen in the malaria parasite *Plasmodium falciparum*(恶性疟原虫疟疾). Here, the reversed action of complex II as an oxidase is important in regenerating ubiquinol, which the parasite uses in an unusual form of pyrimidine(嘧啶) biosynthesis.

(3) Electron transfer flavoprotein - Q oxidoreductase 电子转移黄素蛋白-Q 氧化还原酶

Electron transfer flavoprotein-ubiquinone oxidoreductase (ETF - Q oxidoreductase), also known as *electron transferring-flavoprotein dehydrogenase*, is a third entry point to the electron transport chain. It is an enzyme that accepts electrons from electron-transferring flavoprotein in the mitochondrial matrix, and uses these electrons to reduce ubiquinone. This enzyme contains a flavin and a [4Fe - 4S] cluster, but, unlike the other respiratory complexes, it attaches to the surface of the membrane and does not cross the lipid bilayer.

电子转移黄素蛋白 Q 氧化还原酶的介绍。

$$ETF_{red} + Q \longrightarrow ETF_{ox} + QH_2$$

In mammals, this metabolic pathway is important in β oxidation of fatty acids and catabolism of amino acids and choline(胆碱), as it accepts electrons from multiple acetyl-CoA dehydrogenases. In plants, ETF - Q oxidoreductase is also important in the metabolic responses that allow survival in extended periods of darkness.

在哺乳动物体内，这个代谢途径对脂肪酸的β氧化和氨基酸、胆碱的分解代谢非常重要。

(4) Q-cytochrome C oxidoreductase (complex III) Q-细胞色素 C 氧化还原酶(复合体 III)

Q-cytochrome C oxidoreductase is also known as *cytochrome C reductase*, *cytochrome bc_1 complex*, or simply *complex III*. In mammals, this enzyme is a dimer(二聚体), with each subunit complex containing 11 protein subunits, an [2Fe - 2S] iron-sulfur cluster and three cytochromes: one cytochrome c_1 and two b cytochromes. A cytochrome is a kind of electron-transferring protein that contains at least one heme group. The iron atoms inside complex III's heme groups alternate between a

哺乳动物体内复合体 III 的结构。

reduced ferrous (+2) and oxidized ferric (+3) state as the electrons are transferred through the protein.

The reaction catalyzed by complex III is the oxidation of one molecule of ubiquinol and the reduction of two molecules of cytochrome C, a heme protein loosely associated with the mitochondrion. Unlike coenzyme Q, which carries two electrons, cytochrome C carries only one electron.

$$QH_2 + 2Cyt\ c_{ox} + 2H^+_{matrix} \longrightarrow Q + 2Cyt\ c_{red} + 4H^+_{intermembrane}$$

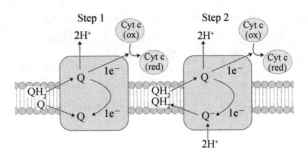

The two electron transfer steps in complex III: Q-cytochrome c oxidoreductase. After each step, Q (in the upper part of the figure) leaves the enzyme.

As only one of the electrons can be transferred from the QH_2 donor to a cytochrome C acceptor at a time, the reaction mechanism of complex III is more elaborate(复杂) than those of the other respiratory complexes, and occurs in two steps called the Q cycle(Q 循环). In the first step, the enzyme binds three substrates, first, QH_2, which is then oxidized, with one electron being passed to the second substrate, cytochrome c. The two protons released from QH_2 pass into the intermembrane space. The third substrate is Q, which accepts the second electron from the QH_2 and is reduced to Q^-, which is the ubisemiquinone(泛半醌) free radical(自由基). The first two substrates are released, but this ubisemiquinone intermediate remains bound. In the second step, a second molecule of QH_2 is bound and again passes its first electron to a cytochrome c acceptor. The second electron is passed to the bound ubisemiquinone, reducing it to QH$_2$ as it gains two protons from the mitochondrial matrix. This QH$_2$ is then released from the enzyme.

As coenzyme Q is reduced to ubiquinol(泛醇) on the inner side of the membrane and oxidized to ubiquinone(泛醌) on the other, a net transfer of protons across the membrane occurs, adding to the proton gradient. The rather complex two-step mechanism by which this occurs is important, as it increases the efficiency of proton transfer. If, instead of the Q cycle, one molecule of QH$_2$ were used to directly reduce two molecules of cytochrome C, the efficiency would be halved(减半), with only one proton transferred per cytochrome C reduced.

Q循环的结果。

Q循环缺陷的后果。

(5) Cytochrome C oxidase(complex IV) 细胞色素 C 氧化酶（复合体 IV）

Cytochrome C oxidase, also known as *complex IV*, is the final protein complex in the electron transport chain. The mammalian(哺乳动物) enzyme has an extremely complicated structure and contains 13 subunits, two heme groups, as well as multiple metal ion cofactors — in all three atoms of copper, one of magnesium(镁) and one of zinc(锌).

细胞色素 C 氧化酶在电子传递链上的作用。

This enzyme mediates the final reaction in the electron transport chain and transfers electrons to oxygen, while pumping protons across the membrane. The final electron acceptor oxygen, which is also called the *terminal electron acceptor*, is reduced to water in this step. Both the direct pumping of protons and the consumption of matrix protons(电子) in the reduction of oxygen contribute to the proton gradient (梯度). The reaction catalyzed is the oxidation(氧化) of cytochrome C and the reduction(还原) of oxygen:

与细胞色素 C 相关的化学反应。

$$4Cyt\ c_{red} + O_2 + 8H^+_{matrix} \longrightarrow 4Cyt\ c_{ox} + 2H_2O + 4H^+_{intermembrane}$$

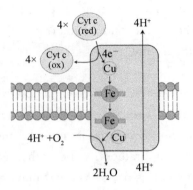

Complex IV: cytochrome c oxidase.

5.3.3 Alternative reductases and oxidases 替代还原酶和氧化酶

Many eukaryotic(真核的) organisms have electron transport chains that differ from the much-studied mammalian(哺乳动物) enzymes described above. For example, plants have alternative NADH oxidases, which oxidize NADH in the cytosol rather than in the mitochondrial matrix(线粒体基质), and pass these electrons to the ubiquinone pool. These enzymes do not transport protons, and, therefore, reduce ubiquinone(泛醌) without altering the electrochemical gradient across the inner membrane.

许多真核生物的电子传递链与上述哺乳动物并不一样。

Another example of a divergent electron transport chain is the alternative oxidase, which is found in plants, as well as some fungi(真菌), protists(原生生物), and possibly some animals. This enzyme transfers electrons directly from ubiquinol to oxygen.

The electron transport pathways produced by these alternative NADH and ubiquinone oxidases have lower ATP yields than the full pathway. The advantages produced by a shortened pathway are not entirely clear. However, the alternative oxidase is produced in response to stresses such as cold, reactive oxygen species(活性氧), and infection by pathogens(病原体), as well as other factors that inhibit the full electron transport chain. Alternative pathways might, therefore, enhance an organisms' resistance to injury, by reducing oxidative stress.

NADH替代物和泛醌氧化酶形成的电子传播途径,所产生的ATP比完整电子传递链要少。缩短路径产生的优势还不完全清楚,但能通过减少氧化胁迫来提高机体的损伤抵制力。

5.3.4 Organization of complexes 复合物的组装

The original model for how the respiratory chain complexes are organized was that they diffuse freely and independently in the mitochondrial membrane. However, recent data suggest that the complexes might form higher order structure(高层次结构) called super complexes or "respirasomes". In this model, the various complexes exist as organized sets of interacting enzymes(相互作用的酶). These associations might allow channeling(通道作用) of substrates between the various enzyme complexes, increasing the rate and efficiency of electron transfer. Within such mammalian supercomplexes(超分子), some components would be present in higher amounts than others, with some data suggesting a ratio between complexes I/II/III/IV and the ATP synthase of approximately 1 : 1 : 3 : 7 : 4. However, the debate over this supercomplex hypothesis is not completely resolved, as some data do not appear to fit with this model.

5.3.5 Prokaryotic electron transport chains 原核生物电子传递链

In contrast to the general similarity in structure and function of the electron transport chains in eukaryotes, bacteria and archaea(古生菌) possess a large variety of electron-transfer enzymes. These use an equally wide set of chemicals as substrates. In common with eukaryotes, prokaryotic electron transport uses the energy released from the oxidation of a substrate to pump ions across a membrane and generate an electrochemical gradient. In the bacteria, oxidative phosphorylation in *Escherichia coli*(大肠杆菌) is understood in most detail, while archaeal systems are at present poorly understood.

The main difference between eukaryotic and prokaryotic oxidative phosphorylation is that bacteria and archaea(古细菌) use many different substances to donate(提供) or accept(接受) electrons. This allows prokaryotes to grow under a wide variety

大肠杆菌的氧化磷酸化。

of environmental conditions. In *E. coli*, for example, oxidative phosphorylation can be driven by a large number of pairs of reducing agents(还原剂) and oxidizing agents(氧化剂), which are listed below. The midpoint potential(中点电位) of a chemical measures how much energy is released when it is oxidized or reduced, with reducing agents having negative potentials(负电势) and oxidizing agents positive potentials(正电势).

大肠杆菌中的呼吸酶和底物。

Respiratory enzymes and substrates in *E. coli*.

Respiratory enzyme	Redox pair	Midpoint potential (Volts)
Formate dehydrogenase	Bicarbonate/Formate	−0.43
Hydrogenase	Proton/Hydrogen	−0.42
NADH dehydrogenase	NAD^+/NADH	−0.32
Glycerol-3-phosphate dehydrogenase	DHAP/Gly-3-P	−0.19
Pyruvate oxidase	Acetate+Carbon dioxide/Pyruvate	—
Lactate dehydrogenase	Pyruvate/Lactate	−0.19
D-amino acid dehydrogenase	2-oxoacid + ammonia/*D*-amino acid	—
Glucose dehydrogenase	Gluconate/Glucose	−0.14
Succinate dehydrogenase	Fumarate/Succinate	+0.03
Ubiquinol oxidase	Oxygen/Water	+0.82
Nitrate reductase	Nitrate/Nitrite	+0.42
Nitrite reductase	Nitrite/Ammonia	+0.36
Dimethyl sulfoxide reductase	DMSO/DMS	+0.16
Trimethylamine *N*-oxide reductase	TMAO/TMA	+0.13
Fumarate reductase	Fumarate/Succinate	+0.03

As shown above, *E. coli* can grow with reducing agents such as formate(甲酸盐), hydrogen, or lactate as electron donors, and nitrate(硝酸盐), DMSO(二甲基亚砜), or oxygen as acceptors. The larger the difference in midpoint potential between an oxidizing and reducing agent, the more energy is released when they react. Out of these compounds, the succinate/fumarate(琥珀酸/延胡索酸) pair is unusual, as its midpoint potential is close to zero. Succinate can therefore be oxidized to fumarate if a strong oxidizing agent such as oxygen is available, or fumarate can be reduced to succinate using a strong reducing agent such as oxygen is available, or fumarate can be reduced to succinate using a strong reducing agent such as formate. These alternative reactions are catalyzed by succinate dehydrogenase(琥珀酸脱氢酶) and fumarate reductase(延胡索酸还原酶), respectively.

琥珀酸/延胡索酸反应对在原核生物氧化磷酸化中的特殊性,它的电位中点接近0。

Some prokaryotes use redox pairs(氧化还原对) that have only a small difference in midpoint potential. For example, nitrifying bacteria(硝化细菌) such as *Nitrobacter*(硝化细菌属) oxidize nitrite(亚硝酸盐) to nitrate(硝酸盐), donating the electrons to oxygen. The small amount of energy released in this reaction is enough to pump protons and generate ATP, but not enough to produce NADH or NADPH directly for use in anabolism(合成代谢). This problem is solved by using a nitrite oxidoreductase to produce enough proton-motive force to run part of the electron transport chain in reverse, causing complex I to generate NADH.

一些原核生物对氧化还原对的利用。

亚硝酸氧化还原酶产生足够的质子动力来反向驱动部分电子传递链,促使complex I生成NADH。

Prokaryotes control their use of these electron donors and acceptors by varying which enzymes are produced, in response to environmental conditions. This flexibility is possible because different oxidases and reductases use the same ubiquinone pool. This allows many combinations(组合) of enzymes to function together, linked by the common ubiquinol intermediate. These respiratory chains therefore have a modular(模块化的) design, with easily interchangeable(可互换的) sets of enzyme systems. In addition to this metabolic diversity, prokaryotes also possess a range of isozymes—different enzymes that catalyze the same reaction. For example, in *E. coli*, there are two different types of ubiquinol oxidase using oxygen as an electron acceptor. Under highly aerobic conditions, the cell uses an oxidase with a low affinity for oxygen that can transport two protons per

原核生物通过不同的酶来控制对电子供体和受体的利用,来应答环境因素。

electron. However, if levels of oxygen fall, they switch to an oxidase that transfers only one proton per electron, but has a high affinity for oxygen.

5.4 ATP synthase (complex V) ATP 合酶(复合体 V)

ATP 合酶是氧化磷酸化的最后一个酶，在原核生物和真核生物中的功能相同。

ATP synthase, also called *complex V*, is the final enzyme in the oxidative phosphorylation pathway. This enzyme is found in all forms of life and functions in the same way in both prokaryotes and eukaryotes. The enzyme uses the energy stored in a proton gradient(质子梯度) across a membrane to drive the synthesis of ATP from ADP and phosphate (P_i). Estimates(估算) of the number of protons required to synthesize one ATP have ranged from three to four, with some suggesting cells can vary this ratio, to suit different conditions.

$$ADP + P_i + 4H^+_{intermembrane} \rightleftharpoons ATP + H_2O + 4H^+_{matrix}$$

可以通过改变质子动力势力来转移该磷酸化反应的平衡。

This phosphorylation reaction is an equilibrium(均衡), which can be shifted by altering the proton-motive force. In the absence of a proton—motive force, the ATP synthase reaction will run from right to left, hydrolyzing ATP and pumping protons out of the matrix across the membrane. However, when the proton-motive force is high, the reaction is forced to run in the opposite direction; it proceeds from left to right, allowing protons to flow down their concentration gradient and turning ADP into ATP. Indeed, in the closely related(与密切相关) vacuolar type(空泡的类型) H^+-ATPases, the same reaction is used to acidify(酸化) cellular compartments(细胞区室), by pumping protons and hydrolysing ATP.

ATP 合酶的结构及其功能。

ATP synthase is a massive protein complex with a mushroom-like shape(蘑菇型的形状). The mammalian enzyme complex contains 16 subunits and has a mass of approximately 600 kilodaltons(千道尔顿). The portion embedded within(嵌

入) the membrane is called F_o and contains a ring of c subunits and the proton channel. The stalk(杆) and the ball-shaped headpiece(球形头部) is called F_1 and is the site of ATP synthesis. The ball-shaped complex at the end of the F_1 portion contains six proteins of two different kinds (three α subunits and three β subunits), whereas the "stalk" consists of one protein: the γ subunit, with the tip of the stalk extending into the ball of α and β subunits. Both the α and β subunits bind nucleotides(核苷酸), but only the β subunits catalyze the ATP synthesis reaction. Reaching along the side of the F_1 portion and back into the membrane is a long rod-like subunit that anchors the α and β subunits into the base of the enzyme.

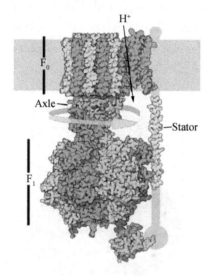

ATP synthase.

As protons cross the membrane through the channel in the base of ATP synthase, the F_o proton-driven motor rotates(电机旋转). Rotation might be caused by changes in the ionization (离子化) of amino acids in the ring of c subunits causing electrostatic interactions(静电相互作用) that propel the ring of c subunits past the proton channel. This rotating ring(旋转环) in turn drives the rotation of the central axle(中心轴)(the γ subunit stalk) within the α and β subunits. The α and β subunits are prevented from rotating themselves by the side-arm, which

当质子通过 ATP 合酶的通道穿过膜层，F_O 质子驱动电机开始旋转。

γ、α、β 亚基的作用。

acts as a stator. This movement of the tip of the γ subunit within the ball of α and β subunits provides the energy for the active sites in the β subunits to undergo a cycle of movements that produces and then releases ATP.

This ATP synthesis reaction is called the *binding change mechanism*(结合变构机制) and involves the active site of a β subunit cycling between three states. In the "open" state, ADP and phosphate enter the active site (shown in brown in the diagram). The protein then closes up around the molecules and binds them loosely — the "loose" state(松散状态). The enzyme then changes shape again and forces these molecules together, with the active site in the resulting "tight" state(紧密状态) binding the newly produced ATP molecule with very high affinity(高度亲和力). Finally, the active site cycles back to the open state, releasing ATP and binding more ADP and phosphate, ready for the next cycle.

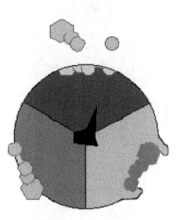

Mechanism of ATP synthase.

In some bacteria and archaea, ATP synthesis is driven by the movement of sodium ions(钠离子) through the cell membrane, rather than the movement of protons. Archaea such as *Methanococcus*(甲烷球菌) also contain the A_1A_o synthase, a form of the enzyme that contains additional proteins with little similarity in sequence to other bacterial and eukaryotic ATP synthase subunits. It is possible that, in some species, the

A_1A_o form of the enzyme is a specialized sodium-driven ATP synthase, but this might not be true in all cases.

5.5 Oxidative stress and Inhibitors 氧化应激与抑制剂

5.5.1 Reactive oxygen species 活性氧类

Molecular oxygen is an ideal terminal electron acceptor because it is a strong oxidizing agent. The reduction of oxygen does involve potentially harmful intermediates. Although the transfer of four electrons and four protons reduces oxygen to water, which is harmless, transfer of one or two electrons produces superoxide(过氧化物) or peroxide anions(过氧化阴离子), which are dangerously reactive.

氧分子是理想的电子终端接受体,因为它是强氧化剂。氧的还原反应包含有害的中间产物。

$$O_2 \xrightarrow{e^-} O_2^- \xrightarrow{e^-} O_2^{2-}$$
$$\quad\quad\quad\;\text{Superoxide}\quad\text{Peroxide}$$

关于氧的化学反应。

These reactive oxygen species and their reaction products, such as the hydroxyl radical(羟基), are very harmful to cells, as they oxidize proteins and cause mutations(突变) in DNA. This cellular damage might contribute to disease and is proposed as one cause of aging.

活性氧类及其反应产物的危害。

The cytochrome C oxidase complex is highly efficient at reducing oxygen to water, and it releases very few partly reduced intermediates; however small amounts of superoxide anion and peroxide are produced by the electron transport chain. Particularly important is the reduction of coenzyme Q in complex III, as a highly reactive ubisemiquinone free radical is formed as an intermediate in the Q cycle. This unstable species can lead to electron "leakage"(渗透) when electrons transfer directly to oxygen, forming superoxide. As the production of reactive oxygen species by these proton-pumping complexes is greatest at high membrane potentials, it has been proposed that

细胞色素C氧化酶复合物的作用。

辅酶Q自由基可以形成Q循环的中间产物。

膜电位的变化。

mitochondria regulate their activity to maintain the membrane potential within a narrow range that balances ATP production against oxidant generation. For instance, oxidants can activate uncoupling proteins that reduce membrane potential(膜电位).

为了抵消活性氧,细胞含有大量的抗氧化系统。抗氧化系统的组成。

To counteract(抵消) these reactive oxygen species, cells contain numerous antioxidant systems(抗氧化系统), including antioxidant vitamins such as vitamin C and vitamin E, and antioxidant enzymes such as superoxide dismutase(超氧化物歧化酶), catalase(过氧化氢酶), and peroxidases(氧化物酶), which detoxify the reactive species, limiting damage to the cell.

5.5.2　Inhibitors 抑制剂

几种众所周知的药物和毒物对氧化磷酸化的抑制。

抑制物产生的危害和例子。

There are several well-known drugs and toxins that inhibit oxidative phosphorylation. Although any one of these toxins inhibits only one enzyme in the electron transport chain, inhibition of any step in this process will halt the rest of the process. For example, if oligomycin(寡霉素) inhibits ATP synthase, protons cannot pass back into the mitochondrion. As a result, the proton pumps are unable to operate, as the gradient becomes too strong for them to overcome. NADH is then no longer oxidized and the citric acid cycle ceases to operate because the concentration of NAD^+ falls below the concentration that these enzymes can use.

氧化磷酸化中不是所有的抑制物都是有毒的,有的是解偶联剂。

Not all inhibitors of oxidative phosphorylation are toxins(毒素). In brown adipose tissue(脂肪组织), regulated proton channels called uncoupling proteins(解偶联蛋白) can uncouple respiration from ATP synthesis. This rapid respiration produces heat, and is particularly important as a way of maintaining body temperature for hibernating animals, although these proteins may also have a more general function in cells' responses to stress.

Compounds	Use	Effect on oxidative phosphorylation
Cyanide Carbon monoxide Azide Hydrogen sulfide	Poisons	Inhibit the electron transport chain by binding more strongly than oxygen to the Fe—Cu center in cytochrome C oxidase, preventing the reduction of oxygen.
Oligomycin	Antibiotic	Inhibits ATP synthase by blocking the flow of protons through the F_0 subunit.
CCCP 2,4-Dinitrophenol	Poisons	Ionophores that disrupt the proton gradient by carrying protons across a membrane. This ionophore uncouples proton pumping from ATP synthesis because it carries protons across the inner mitochondrial membrane.
Rotenone	Pesticide	Prevents the transfer of electrons from complex I to ubiquinone by blocking the ubiquinone-binding site.
Malonate and oxaloacetate	Poisons	Competitive inhibitors of succinate dehydrogenase (complex II)
Antimycin A	Piscicide	Binds to the Q_i site of cytochrome C reductase, thereby inhibiting the oxidation of ubiquinol

【Key Words】

oxidative phosphorylation 氧化磷酸化	pyruvate dehydrogenase complex 丙酮酸脱氢酶复合体
electron-transport chain 电子传递链	NADH-Q oxidoreductase NADH－Q 氧化还原酶
iron-sulfur protein 铁硫蛋白	succinate－Q reductase 琥珀酸辅酶 Q 还原酶
lavoprotein 黄素蛋白	Q-cytochrome C oxidoreductase Q－细胞色素 C 氧化还原酶
ubiquinone 泛醌	cytochrome C oxidase 细胞色素 C 氧化酶
ATP synthase ATP 合酶	glycerol 3-phosphate shuttle 磷酸甘油穿梭
superoxide dismutase 超氧化物歧化酶	malate－aspartate shuttle 苹果酸-天冬氨酸穿梭
flavin mononucleotide (FMN) 黄素单核苷酸	redox reactions 氧化还原反应

Questions

1. Briefly define the Complex I, II, III, IV and V.
2. Write down the location of specific ATP-producing sites in the electron transport chain.
3. Briefly define the following terms:
    (a) Electron transport chain.
    (b) Oxidative phosphorylation and P/O ratio.
    (c) Redox reaction.

4. A biological redox reaction always involves:

   (a) A loss of electron.

   (b) A gain of electron.

   (c) A reducing agent.

   (d) An oxidizing agent.

   (e) All of the above.

References

[1] McKee Trudy, R McKee James. Biochemistry: An Introduction (Second Edition)[M]. New York: McGraw-Hill Companies, 1999.

[2] Nelson D L, Cox M M. Lehninger Principles of Biochemistry (Third Edition)[M]. Derbyshire: Worth Publishers, 2000.

[3] Ding W, Jia H T. Biochemistry (First Edition)[M]. Beijing: Higher Education Press, 2012.

[4] 郑集, 陈钧辉. 普通生物化学(第四版)[M]. 北京: 高等教育出版社, 2007.

[5] Berg J M, Tymoczko J L, Styer L. Biochemistry (Seventh Edition)[M]. New York: W. H. Freeman and Company, 2012.

Part III
Informational Macromolecules

信息大分子

Chapter 1　DNA synthesis and repair
DNA 合成与修复

1.1　DNA replication DNA 复制

DNA replication is a biological process that occurs in all living organisms(生物体) and copies their DNA; it is the basis for biological inheritance(生物遗传). The process starts with one double-stranded DNA molecule(双链 DNA 分子) and produces two identical(完全相同的) copies of the molecule. Each strand of the original double-stranded DNA molecule serves as template(模板) for the production of the complementary strand(互补链). Cellular proofreading(校对) and error checking mechanisms(错误检查机制) ensure near perfect fidelity(忠实性) for DNA replication.

> DNA 分子复制的概念以及原理。

In a cell, DNA replication begins at specific locations(特定的位置) in the genome(基因组), called "origins". Unwinding of DNA(解旋的 DNA) at the origin(起始点), and synthesis of new strands, forms a replication fork(复制叉). In addition to DNA polymerase(DNA 聚合酶), the enzyme that synthesizes(合成) the new DNA by adding nucleotides(核苷酸) matched to the template strand(模板链), a number of other proteins are associated with the fork and assist in the initiation(起始) and continuation of DNA synthesis.

> DNA 分子复制的过程及特点。

DNA replication can also be performed in vitro(体外) (artificially, outside a cell). DNA polymerases(DNA 聚合酶), isolated from cells, and artificial DNA primers(人工合成的引物) are used to initiate DNA synthesis at known sequences(已知序列) in a template molecule. The polymerase chain reaction (PCR，聚合酶链式反应), a common laboratory technique,

> DNA 复制的应用。

employs such artificial synthesis in a cyclic manner(循环方式) to amplify(放大) a specific target DNA fragment(特定目标 DNA 片段) from a pool of DNA.

DNA replication. The double helix is unwound and each strand acts as a template for the next strand. Bases are matched to synthesize the new partner strands.

(1) DNA structure DNA 结构

DNA分子的双螺旋结构与分子组成。

DNA usually exists as a double-stranded structure, with both strands coiled together to form the characteristic double-helix(双螺旋). Each single strand of DNA is a chain of four types of nucleotides(核苷酸) having the bases(碱基): adenine (腺嘌呤), cytosine(胞嘧啶), guanine(鸟嘌呤), and thymine (胸腺嘧啶). A nucleotide is a mono-, di-, or triphosphate deoxyribonucleoside(单磷酸脱氧核苷、二磷酸脱氧核苷或者三磷酸脱氧核苷); that is, a deoxyribose sugar(脱氧核糖) is attached to one, two, or three phosphates. Chemical interaction of these nucleotides forms phosphodiester(磷酸二酯键) linkages, creating the phosphate-deoxyribose backbone(磷酸-脱氧核糖骨架) of the DNA double helix with the bases pointing inward. Nucleotides (bases) are matched between strands through hydrogen bonds(氢键) to form base pairs(碱基对).

Adenine(腺嘌呤) pairs with thymine(胸腺嘧啶), and cytosine(胞嘧啶) pairs with guanine(鸟嘌呤).

DNA strands have a directionality(方向性), and the different ends of a single strand are called the "3′ (three-prime) end" and the "5′ (five-prime) end". These terms refer to the carbon atom in deoxyribose(脱氧核糖) to which the next phosphate in the chain attaches. In addition to being complementary(互补的), the two strands of DNA are antiparallel(反相平行的): They are orientated in opposite directions. This directionality has consequences in DNA synthesis(DNA 合成), because DNA polymerase(DNA 聚合酶) can synthesize DNA in only one direction by adding nucleotides(核苷酸) to the 3′ end of a DNA strand.

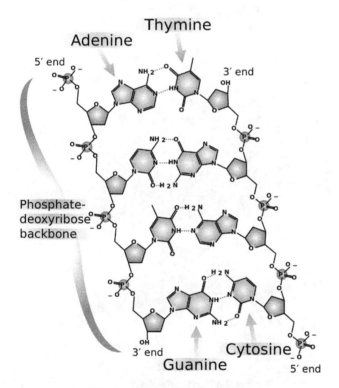

The chemical structure of DNA.

The pairing of bases in DNA through hydrogen bonding(氢键) means that the information contained within each strand is redundant(多余的). The nucleotides on a single strand can be used to reconstruct nucleotides on a newly synthesized(新合成的) partner strand.

(2) DNA polymerase DNA 聚合酶

DNA polymerases are a family of enzymes that carry out all forms of DNA replication(DNA 复制). However, a DNA polymerase can only extend an existing DNA(现有的 DNA 链) strand paired with a template strand(模板链); it cannot begin the synthesis of a new strand. To begin synthesis, a short fragment(短片段) of DNA or RNA, called a primer(引物), must be created and paired with the template DNA strand(模板链).

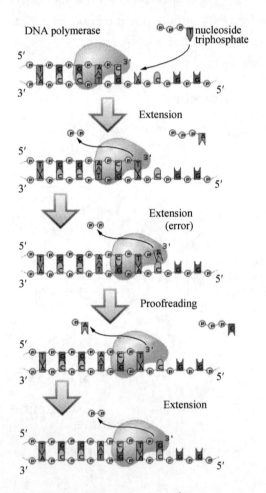

DNA polymerases adds nucleotides to the 3′ end of a strand of DNA. If a mismatch is accidentally incorporated, the polymerase is inhibited from further extension. Proofreading removes the mismatched nucleotide and extension continues.

DNA polymerase then synthesizes a new strand of DNA by extending(延伸) the 3' end of an existing nucleotide chain, adding new nucleotides matched to the template strand one at a time via the creation of phosphodiester bonds(磷酸二酯键). The energy for this process of DNA polymerization(DNA 聚合) comes from two of the three total phosphates(磷酸) attached to each unincorporated base(未结合的碱基). Free bases(游离碱基) with their attached phosphate groups are called nucleoside triphosphates(三磷酸核苷). When a nucleotide(核苷酸) is being added to a growing DNA strand, two of the phosphates are removed and the energy produced creates a phosphodiester bond(磷酸二酯键) that attaches the remaining phosphate to the growing chain. The energetics of this process also help explain the directionality(方向性) of synthesis, if DNA were synthesized in the 3' to 5' direction, the energy for the process would come from the 5' end of the growing strand rather than from free nucleotides.

1.1.1 Replication process 复制过程

(1) Origins 起点

For a cell to divide(细胞分裂), it must first replicate(复制) its DNA. This process is initiated(开始) at particular points in the DNA, known as "origins"(起点), which are targeted by proteins(蛋白质) that separate the two strands and initiate DNA synthesis. Origins contain DNA sequences(DNA 序列) recognized by replication initiator proteins (e.g., DnaA in *E. coli*(大肠杆菌) and the origin recognition complex(起点识别复合物) in yeast(酵母菌)). These initiator proteins recruit(招募) other proteins to separate the two strands and initiate replication forks(复制叉).

Initiator proteins recruit other proteins to separate the DNA strands at the origin, forming a bubble. Origins tend to be "AT-rich" (rich in adenine and thymine bases, 丰富的腺嘌呤和胸腺嘧啶) to assist this process, because A-T base pairs have two hydrogen bonds (氢键) (rather than the three formed in a C-G pair) —in general, strands rich in these nucleotides(核

苷酸) are easier to separate because a greater number of hydrogen bonds requires more energy to break them. Once strands are separated, RNA primers are created on the template strands(模板链). To be more specific(具体的), the leading strand(前导链) receives one RNA primer per active(激活) origin of replication while the lagging strand(后随链) receives several; these several fragments(几个片段) of RNA primers found on the lagging strand of DNA are called Okazaki fragments(冈崎片段), named after their discoverer. DNA polymerase(DNA 聚合酶) extends the leading strand(前导链) in one continuous motion and the lagging strand in a discontinuous(不连续的) motion (due to the Okazaki fragments). RNase(核糖核酸酶) removes the RNA fragments used to initiate replication by DNA Polymerase(DNA 聚合酶), and another DNA Polymerase enters to fill the gaps(缺口). When this is complete, a single nick(缺口) on the leading strand and several nicks on the lagging strand can be found. Ligase(连接酶) works to fill these nicks in, thus completing the newly replicated DNA molecule(DNA 分子).

As DNA synthesis continues, the original DNA strands continue to unwind(解旋) on each side of the bubble, forming a replication fork with two prongs(叉子齿). In bacteria(细菌), which have a single origin of replication on their circular chromosome(环形染色体), this process eventually creates a "theta structure"(resembling the Greek letter(希腊字母) *theta*: θ). In contrast(相反), eukaryotes(真核生物) have longer linear chromosomes(较长的线性染色体) and initiate replication at multiple origins(多起点开始复制) within these.

（2）**Replication fork 复制叉**

The replication fork is a structure that forms within the nucleus(细胞核) during DNA replication. It is created by helicases(解旋酶), which break the hydrogen bonds(氢键) holding the two DNA strands together. The resulting structure has two branching "prongs"(分支), each one made up of a single strand of DNA. These two strands serve as the template for the leading and lagging strands(前导链和后随链), which will be created as DNA polymerase（DNA 聚合酶）matches complementary（互补的）nucleotides to the templates; The templates may be properly referred to as the leading strand template and the lagging strand template.

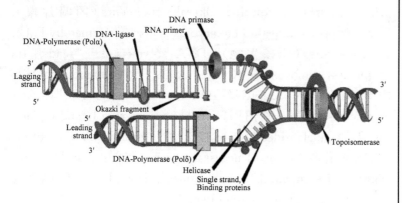

Many enzymes are involved in the DNA replication fork.

图示 DNA 复制叉上的多种酶的参与。

Leading strand 前导链

The leading strand is the template strand of the DNA double helix（双螺旋结构）so that the replication fork（复制叉）moves along it in the 3′ to 5′ direction. This allows the new strand synthesized complementary（互补的）to it to be synthesized 5′ to 3′ in the same direction as the movement of the replication fork.

前导链上 DNA 的复制过程和特点。

On the leading strand, a polymerase（聚合酶）"reads" the DNA and adds nucleotides（核苷酸）to it continuously. This polymerase is DNA polymerase III（DNA 聚合酶 III）in prokaryotes（原核生物）and presumably Pol ε in eukaryotes（真核生物）.

原核生物和真核生物复制时 DNA 聚合酶的不同。

Lagging strand 后随链

The lagging strand is the strand of the template DNA double helix that is oriented（标定方向的）so that the replication fork moves along it in a 5′ to 3′ manner. Because of its orientation（方向），opposite to the working orientation of DNA polymerase III, which moves on a template in a 3′ to 5′ manner, replication of the lagging strand is more complicated

后随链上的 DNA 复制过程及特点。

than that of the leading strand. On the lagging strand(后随链), primase "reads" the DNA and adds RNA to it in short, separated segments. In eukaryotes(真核生物), primase is intrinsic to Pol α. DNA polymerase III or Pol δ lengthens the primed segments, forming Okazaki fragments(冈崎片段). Primer(引物) removal in eukaryotes is also performed by Pol δ. In prokaryotes(原核生物), DNA polymerase I "reads" the fragments, removes the RNA using its flap endonuclease(核酸内切酶) domain. RNA primers are removed by 5'-3' exonuclease(核酸外切酶) activity of polymerase I, and replaces the RNA nucleotides(RNA 核苷酸) with DNA nucleotides(DNA 核苷酸)(this is necessary because RNA and DNA use slightly different kinds of nucleotides). DNA ligase(DNA 连接酶) joins the fragments together.

Dynamics at the replication fork 复制叉动力学

As helicase(解旋酶) unwinds DNA at the replication fork, the DNA ahead is forced to rotate(旋转). This process results in a build-up(形成、积累) of twists in the DNA ahead. This build-up would form a resistance that would eventually halt(停止、终止) the progress of the replication fork. DNA topoisomerases(拓扑异构酶) are enzymes(酶) that solve these physical problems in the coiling of DNA(DNA 螺旋). Topoisomerase I cuts a single backbone(主干) on the DNA, enabling the strands to swivel around(转动) each other to remove the build-up of twists. Topoisomerase II cuts both backbones, enabling one double-stranded DNA to pass through another, thereby removing knots(结) and entanglements(缠绕) that can form within and between DNA molecules.

Bare single-stranded DNA tends to fold back on itself and form secondary structures(二级结构); these structures can interfere(干扰) with the movement of DNA polymerase(DNA 聚合酶). To prevent this, single-strand binding proteins bind to the DNA until a second strand is synthesized(合成), preventing secondary structure(二级结构) formation.

Clamp proteins(钳蛋白) form a sliding clamp around

DNA, helping the DNA polymerase(聚合酶) maintain contact with its template(模板), thereby assisting with processivity(持续合成能力). The inner face of the clamp enables DNA to be threaded through it. Once the polymerase reaches the end of the template or detects(发现) double-stranded DNA, the sliding clamp undergoes a conformational change(构象变化) that releases the DNA polymerase. Clamp-loading proteins are used to initially load the clamp, recognizing the junction(连接) between template and RNA primers(引物).

(3) Termination 终止

Eukaryotes(真核生物) initiate DNA replication at multiple points in the chromosome(染色体), so replication forks meet and terminate at many points in the chromosome; these are not known to be regulated in any particular way. Because eukaryotes have linear(线性的) chromosomes, DNA replication is unable to reach the very end of the chromosomes, but ends at the telomere(端粒) region of repetitive DNA close to the end. This shortens the telomere(端粒) of the daughter DNA strand(子代 DNA 链). This is a normal process in somatic cells(体细胞). As a result, cells can only divide a certain number of times before the DNA loss prevents further division. This is known as the Hayflick(海弗利克) limit. Within the germ cell(生殖细胞) line, which passes DNA to the next generation, telomerase extends the repetitive sequences of the telomere region to prevent degradation. Telomerase can become mistakenly active in somatic cells(体细胞), sometimes leading to cancer formation(癌症的形成).

Because bacteria(细菌) have circular chromosomes(环状染色体), termination of replication occurs when the two replication forks meet each other on the opposite end of the parental chromosome. E. coli(大肠杆菌) regulate this process through the use of termination sequences that, when bound by the Tus protein, enable only one direction of replication fork to pass through. As a result, the replication forks are constrained (约束) to always meet within the termination region of the chromosome.

真核生物复制有多个起点。

海弗利克极限：细胞只能分裂到一定次数。

1.1.2　Polymerase chain reaction 聚合酶链式反应

PCR 技术的原理。

Researchers commonly replicate DNA in vitro(体外) using the polymerase chain reaction (PCR). PCR uses a pair of primers(引物) to span a target region in template DNA, and then polymerizes(使聚合) partner strands in each direction from these primers using a thermostable(热稳定) DNA polymerase. Repeating this process through multiple cycles produces amplification of the targeted DNA region. At the start of each cycle, the mixture of template and primers is heated, separating the newly synthesized molecule and template. Then, as the mixture cools, both of these become templates for annealing of new primers, and the polymerase extends from these. As a result, the number of copies of the target region doubles each round, increasing exponentially(指数地).

1.2　DNA repair DNA 修复

DNA 修复的概念。

导致 DNA 损伤的因素。

DNA 损伤导致的结果。

DNA repair refers to a collection of processes by which a cell identifies and corrects damage to the DNA molecules that encode(编码) its genome(基因组). In human cells, both normal metabolic activities(代谢活动) and environmental factors such as UV light(紫外线) and radiation(辐射) can cause DNA damage, resulting in as many as 1 million individual molecular lesions(损伤) per cell per day. Many of these lesions cause structural damage to the DNA molecule and can alter(改变) or eliminate(消除) the cell's ability to transcribe(转录) the gene that the affected DNA encodes(DNA 编码). Other lesions (损伤) induce potentially(潜在的) harmful mutations(突变) in the cell's genome, which affect the survival of its daughter cells after it undergoes mitosis(有丝分裂). As a consequence, the DNA repair process is constantly active as it responds to damage in the DNA structure. When normal repair processes fail, and when cellular apoptosis(细胞凋亡) does not occur, irreparable (无法挽回的) DNA damage may occur, including double-strand breaks and DNA crosslinkages(交联).

The rate of DNA repair is dependent on many factors, including the cell type, the age of the cell, and the extracellular(细胞外) environment. A cell that has accumulated a large amount of DNA damage, or one that no longer effectively repairs damage incurred to its DNA, can enter one of three possible states:

1. an irreversible(不可逆的) state of dormancy(休眠状态), known as senescence(衰老);

2. cell suicide, also known as apoptosis(细胞凋亡) or programmed cell death(程序性细胞死亡);

3. unregulated cell division(细胞分裂), which can lead to the formation of a tumor(肿瘤) that is cancerous.

The DNA repair ability of a cell is vital to the integrity of its genome and thus to its normal functioning and that of the organism(器官). Many genes that were initially shown to influence life span have turned out to be involved in DNA damage repair and protection. Failure to correct molecular lesions(损伤) in cells that form gametes(配子) can introduce mutations(突变) into the genomes of the offspring(子代) and thus influence the rate of evolution(进化).

1.2.1 DNA damage DNA 损伤

DNA damage, due to environmental factors and normal metabolic(代谢) processes inside the cell, occurs at a rate of 1,000 to 1,000,000 molecular lesions(损伤) per cell per day. While this constitutes only 0.000 165% of the human genome's approximately 6 billion bases (3 billion base pairs), unrepaired lesions in critical genes (such as tumor suppressor genes(肿瘤抑制基因)) can impede(阻止) a cell's ability to carry out its function and appreciably increase the likelihood of tumor formation(肿瘤形成). The vast majority of DNA damage affects the primary structure of the double helix(双螺旋); that is, the bases themselves are chemically modified. These modifications can in turn disrupt the molecules' regular helical structure(螺旋结构) by introducing non-native chemical bonds or bulky adducts(加合物) that do not fit in the standard double helix. Unlike proteins and RNA, DNA usually lacks tertiary structure(三级结构) and therefore damage or disturbance does not occur at that level. DNA is, however, supercoiled(超螺旋结构) and wound around "packaging" proteins called histones(组蛋白) (in eukaryotes(真核生物)), and both superstructures(高级结构) are vulnerable to the effects of DNA damage.

(1) Sources of damage 损伤的来源

DNA damage can be subdivided into two main types:

1. endogenous damage(内源性损伤) such as attack by reactive oxygen(氧) species produced from normal metabolic byproducts(代谢副产物) (spontaneous mutation(自发突变)), especially the process of oxidative deamination(氧化脱氨作用); also includes replication errors.

2. exogenous damage(外源性损伤) caused by external agents such as:

① ultraviolet (UV 200～300 nm) radiation(紫外线辐射) from the sun;

② other radiation frequencies, including X-rays(X-射线) and gamma rays(γ-射线);

③ hydrolysis(水解) or thermal disruption(热裂解);

④ certain plant toxins(某些植物毒素);

⑤ human-made mutagenic chemicals(化学诱变剂), especially aromatic compounds(芳香族化合物) that act as DNA intercalating agents;

⑥ cancer chemotherapy and radiotherapy(癌症化疗和放疗);

⑦ viruses(病毒).

The replication of damaged DNA before cell division(细胞分裂) can lead to the incorporation of wrong bases opposite damaged ones. Daughter cells that inherit(继承) these wrong bases carry mutations(突变) from which the original DNA sequence is unrecoverable, except in the rare case of a back mutation, for example, through gene conversion(基因转换).

(2) Types of damage 损伤类型

There are five main types of damage to DNA due to

endogenous(内源性) cellular processes：

1. Oxidation of bases and generation of DNA strand interruptions from reactive oxygen species.

2. Alkylation(烷基化) of bases (usually methylation(甲基化)), such as formation of 7-methylguanine(7-甲基鸟嘌呤), 1-methyladenine(1-甲基腺嘌呤), 6-O-Methylguanine.

3. Hydrolysis(水解) of bases, such as deamination(脱氨), depurination(脱嘌呤), and depyrimidination(脱嘧啶).

4. "Bulky adduct formation" (i. e., benzo[a]pyrene diol epoxide-dG adduct).

5. Mismatch of bases, due to errors in DNA replication (DNA 复制), in which the wrong DNA base is stitched into place in a newly forming DNA strand, or a DNA base is skipped over or mistakenly inserted.

Damage caused by exogenous(外源性) agents comes in many forms. Some examples are：

1. UV-B light causes crosslinking(交联) between adjacent cytosine and thymine bases(相邻的胞嘧啶和胸腺嘧啶) creating pyrimidine dimers(嘧啶二聚体). This is called direct DNA damage.

2. UV-A light creates mostly free radicals(自由基). The damage caused by free radicals is called indirect DNA damage.

3. Ionizing radiation(电离辐射) such as that created by radioactive decay(衰变) or in cosmic rays(宇宙射线) causes breaks in DNA strands. Low-level ionizing radiation may induce irreparable DNA damage (leading to replicational(复制) and transcriptional(转录) errors needed for neoplasia(肿瘤形成) or may trigger viral interactions(触发病毒的相互作用)) leading to pre-mature aging(衰老) and cancer.

4. Thermal disruption at elevated temperature increases the rate of de-purination(脱嘌呤作用) (loss of purine(嘌呤) bases from the DNA backbone) and single-strand breaks. For example, hydrolytic(水解) depurination(脱嘌呤) is seen in the thermophilic bacteria(嗜热的细菌), which grow in hot springs at 40～80℃. The rate of depurination(脱嘌呤) (300 purine residues per genome per generation) is too high in these species to be repaired by normal repair machinery, hence a possibility

引起 DNA 损伤的外源性损伤剂有多种形式。

of an adaptive response cannot be ruled out.

5. Industrial chemicals such as vinyl chloride(氯乙烯) and hydrogen peroxide(过氧化氢), and environmental chemicals such as polycyclic hydrocarbons(多环芳烃) found in smoke, soot and tar(烟尘和焦油) create a huge diversity of DNA adducts-ethenobases(乙烯基), oxidized bases, alkylated phosphotriesters(烷基磷酸三酯) and crosslinking of DNA just to name a few.

UV(紫外线) damage, alkylation/methylation(烷基化或甲基化), X-ray damage and oxidative(氧化性的) damage are examples of induced damage. Spontaneous(自发性的) damage can include the loss of a base, deamination(去氨基), sugar ring puckering(折叠) and tautomeric(互变异构的) shift.

1.2.2 DNA repair mechanisms DNA 修复机制

Cells cannot function if DNA damage(DNA 损失) corrupts the integrity(完整性) and accessibility of essential information in the genome(基因组) (but cells remain superficially functional when so-called "non-essential" genes are missing or damaged). Depending on the type of damage inflicted on the DNA's double helical structure(DNA 双螺旋结构), a variety of repair strategies have evolved to restore lost information. If possible, cells use the unmodified complementary strand of the DNA (DNA 的互补链) or the sister chromatid(染色单体) as a template to recover the original information. Without access to a template, cells use an error-prone recovery mechanism known as translation(翻译) synthesis(合成) as a last resort.

在没有模板的情况下,细胞会使用易出错的翻译合成恢复机制作为最后手段。

Damage to DNA alters the spatial configuration(空间构象) of the helix(螺旋), and such alterations can be detected by the cell. Once damage is localized, specific DNA repair molecules bind at or near the site of damage, inducing other molecules to bind and form a complex that enables the actual repair to take place.

(1) Direct reversal 直接逆转

Cells are known to eliminate three types of damage to their DNA by chemically reversing it. These mechanisms(机制) do not require a template(模板), since the types of damage they counteract can occur in only one of the four bases. Such direct reversal mechanisms are specific to the type of damage incurred and do not involve breakage of the phosphodiester backbone(磷酸二酯键). The formation of pyrimidine dimers(嘧啶二聚体) upon irradiation with UV light results in an abnormal covalent bond(共价键) between adjacent pyrimidine bases(相邻的嘧啶碱基). The photoreactivation(光复活) process directly reverses this damage by the action of the enzyme photolyase(光裂合酶), whose activation is obligately dependent on energy absorbed from blue/UV light（300～500 nm wavelength）to promote catalysis(催化). Another type of damage, methylation(甲基化) of guanine bases(鸟嘌呤), is directly reversed by the protein methyl guanine methyl transferase(甲基鸟嘌呤甲基转移酶, MGMT), the bacterial(细菌) equivalent of which is called ogt. This is an expensive process because each MGMT molecule can be used only once; that is, the reaction is stoichiometric(化学计算的) rather than catalytic(催化). A generalized response to methylating(甲基化的) agents in bacteria(细菌) is known as the adaptive response and confers a level of resistance to alkylating(烷基化的) agents upon sustained exposure by upregulation of alkylation repair enzymes(烷基化修复酶). The third type of DNA damage reversed by cells is certain methylation(甲基化) of the bases cytosine(胞嘧啶) and adenine(腺嘌呤).

直接逆转机制的特点。

直接逆转的类型：
1. 光复活；
2. 甲基鸟嘌呤甲基转移酶及烷基化修复酶的作用；
3. 胞嘧啶和腺嘌呤的甲基化的逆转。

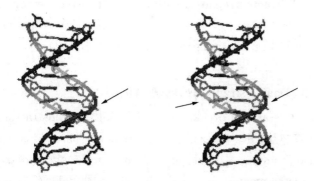

Single-strand and double-strand DNA damage.

(2) Single-strand damage 单链损伤

When only one of the two strands of a double helix(双螺旋) has a defect, the other strand can be used as a template to guide the correction of the damaged strand. In order to repair damage to one of the two paired molecules of DNA, there exist a number of excision repair mechanisms(切除修复机制) that remove the damaged nucleotide(核苷酸) and replace it with an undamaged nucleotide complementary to that found in the undamaged DNA strand.

1. Base excision repair(碱基切除修复, BER), which repairs damage to a single base caused by oxidation(氧化), alkylation(烷基化) hydrolysis(水解), or deamination(去氨基). The damaged base is removed by a DNA glycosylase(DNA 糖基化酶). The "missing tooth" is then recognized by an enzyme called AP endonuclease(脱嘌呤嘧啶核酸内切酶), which cuts the phosphodiester bond(磷酸二酯键). The missing part is then resynthesized by a DNA polymerase(DNA 聚合酶), and a DNA ligase(DNA 连接酶) performs the final nick-sealing step.

2. Nucleotide excision repair(核苷酸切除修复, NER), which recognizes bulky, helix-distorting(螺旋扭曲) lesions(损伤) such as pyrimidine dimers(嘧啶二聚体) and 6,4 photoproducts. A specialized form of NER known as transcription-coupled(转录偶联) repair deploys NER enzymes to genes that are being actively transcribed.

3. Mismatch repair (MMR), which corrects errors of DNA replication(DNA 复制) and recombination that result in mispaired (but undamaged) nucleotides(核苷酸).

(3) Double-strand breaks 双链断裂

Double-strand breaks, in which both strands in the double helix(双螺旋) are severed, are particularly hazardous to the cell because they can lead to genome rearrangements. Three mechanisms exist to repair double-strand breaks (DSBs): non-homologous end joining (NHEJ, 非同源末端连接),

microhomology-mediated end joining（MMEJ，微同源介导的末端连接），and homologous recombination（同源重组）.

In NHEJ，DNA Ligase（DNA 连接酶）IV, a specialized DNA ligase that forms a complex with the cofactor（辅助因子）XRCC4, directly joins the two ends. To guide accurate repair, NHEJ relies on short homologous sequences（同源序列）called microhomologies（微同源）present on the single-stranded tails of the DNA ends to be joined. If these overhangs are compatible, repair is usually accurate. NHEJ can also introduce mutations（突变）during repair. Loss of damaged nucleotides（核苷酸）at the break site can lead to deletions, and joining of nonmatching termini forms translocations（易位）. NHEJ is especially important before the cell has replicated its DNA, since there is no template（模板）available for repair by homologous（同源）recombination. There are "backup" NHEJ pathways in higher eukaryotes（真核生物）. Besides its role as a genome caretaker（基因组看守）, NHEJ is required for joining hairpin-capped double-strand breaks induced during V(D)J recombination, the process that generates diversity in B-cell and T-cell receptors in the vertebrate immune system（脊椎动物的免疫系统）.

> NHEJ 可以作为基因组的守护者，可以使脊椎动物的免疫系统中 B 细胞和 T 细胞受体产生多样性。

Homologous recombination（同源重组）requires the presence of an identical or nearly identical sequence to be used as a template（模板）for repair of the break. The enzymatic machinery responsible for this repair process is nearly identical to the machinery responsible for chromosomal（染色体的）crossover during meiosis（减数分裂）. This pathway allows a damaged chromosome（染色体）to be repaired using a sister chromatid（姐妹染色单体）（available in G2 after DNA replication）or a homologous chromosome（同源染色体）as a template. DSBs caused by the replication machinery attempting to synthesize across a single-strand break or unrepaired lesion cause collapse of the replication fork（复制叉）and are typically repaired by recombination.

> 减数分裂中的酶机制。

> 这种途径允许用一个姐妹染色单体或者同源染色体作为模板。

Topoisomerases（拓扑异构酶）introduce both single-and double-strand breaks in the course of changing the DNA's state

> 拓扑异构酶的作用。

of supercoiling(超螺旋), which is especially common in regions near an open replication fork(复制叉). Such breaks are not considered DNA damage because they are a natural intermediate in the topoisomerase(拓扑异构酶) biochemical mechanism and are immediately repaired by the enzymes that created them.

(4) Translesion synthesis 跨损伤合成

Translesion synthesis (TLS) is a DNA damage tolerance process that allows the DNA replication machinery to replicate past DNA lesions such as thymine dimers(胸腺嘧啶二聚体) or AP sites(嘌呤/嘧啶部位). It involves switching out regular DNA polymerases(DNA 聚合酶) for specialized translesion polymerases, often with larger active sites that can facilitate the insertion of bases(碱基) opposite damaged nucleotides(核苷酸). The polymerase(聚合酶) switching is thought to be mediated by, among other factors, the post-translational modification(修饰) of the replication processivity factor(持续合成因子) PCNA. Translesion synthesis polymerases(聚合酶) often have low fidelity(保真度) on undamaged templates relative to regular polymerases. However, many are extremely efficient at inserting correct bases opposite specific types of damage. From a cellular perspective, risking the introduction of point mutations during translesion synthesis may be preferable to resorting to more drastic mechanisms of DNA repair, which may cause gross chromosomal aberrations(差错) or cell death. In short, the process involves specialized polymerases(聚合酶) either bypassing or repairing lesions(损伤) at locations of stalled(停滞的) DNA replication(复制).

A bypass platform is provided to these polymerases by proliferating cell nuclear antigen (PCNA, 增殖细胞核抗原). Under normal circumstances, PCNA bound to polymerases replicates the DNA. At asite of lesion, PCNA is ubiquitinated (泛素化), or modified, by the RAD6/RAD18 proteins to provide a platform for the specialized polymerases to bypass the lesion(损伤) and resume DNA replication. After translesion synthesis, extension is required. This extension can be carried out by a replicative polymerase(复制的聚合酶) if the TLS is error-free, as in the case of pol η, yet if TLS results in a mismatch, a specialized polymerase is needed to extend it; Pol ζ. Pol ζ is unique in that it can extend terminal mismatches, whereas more processive polymerases cannot. So when a lesion is encountered, the replication fork(复制叉) will stall, PCNA will switch from a processive polymerase to a TLS polymerase to fix the lesion, then PCNA may switch to Pol ζ to extend the mismatch, and last PCNA will switch to the processive polymerase(加工聚合酶) to continue replication.

【Key words】

replication 复制	polymerase chain reaction (PCR) 聚合酶链式反应
polymerases 聚合酶	RNase 核糖核酸酶
ligase 连接酶	leading strands 前导链
template strands 模板链	lagging strands 后随链
replication forks 复制叉	photoreactivation 光复活
Okazaki fragments 冈崎片段	translesion synthesis 跨损伤合成

Questions

1. Describe the terms of replication fork, Okazaki fragments, leading strand, and lagging strand.

2. DNA replicates through a process which is called (   ).
   A. Dispersive replication
   B. Semi-dispersive replication
   C. Conservative replication
   D. Semiconservative replication

3. The enzyme which bonds new nucleotides to those existing on the original DNA strand is called (   ).
   A. DNA polymerase   B. DNA amylase
   C. DNA ligase   D. DNA helicase

References

[1] McKee Trudy, R McKee James. Biochemistry: An Introduction (Second Edition)[M]. New York: McGraw-Hill Companies, 1999.

[2] Nelson D L, Cox M M. Lehninger Principles of Biochemistry (Third Edition)[M].

Derbyshire: Worth Publishers, 2000.

[3] Ding W, Jia H T. Biochemistry (First Edition)[M]. Beijing: Higher Education Press, 2012.

[4] 郑集, 陈钧辉. 普通生物化学(第四版)[M]. 北京:高等教育出版社, 2007.

[5] 王艳萍. 生物化学[M]. 北京:中国轻工业出版社, 2013.

Chapter 2　RNA biosynthesis and processing
RNA 生物合成与加工

2.1　Transcription 转录

Transcription is the process of creating a complementary (互补的) RNA copy of a sequence of DNA. Both RNA and DNA are nucleic acids, which use base pairs of nucleotides(核苷酸) as a complementary language that can be converted back and forth from DNA to RNA by the action of the correct enzymes. During transcription, a DNA sequence is read by RNA polymerase（RNA 聚合酶）, which produces a complementary, antiparallel(反向平行的) RNA strand. As opposed to DNA replication（复制）, transcription results in an RNA complement that includes uracil (U) in all instances where thymine (T, 胸腺嘧啶) would have occurred in a DNA complement.

转录的含义。

转录的过程。

Transcription can be explained easily in 4 or 5 steps, each moving like a wave along the DNA.

1. Helicase（解螺旋酶）unwinds/"unzips" the DNA by breaking the hydrogen bonds（氢键）between complementary nucleotides(核苷酸).

2. RNA nucleotides are paired with complementary DNA bases.

3. RNA sugar-phosphate backbone(糖-磷酸酯骨架) forms with assistance from RNA polymerase(RNA 聚合酶).

4. Hydrogen bonds of the untwisted(松开的) RNA+DNA helix break, freeing the newly synthesized RNA strand.

5. If the cell has a nucleus, the RNA is further processed (addition of a 3′ poly-A tail and a 5′ cap) and exits through to the cytoplasm(细胞浆) through the nuclear pore complex(核孔复合物).

用 4 或 5 个步骤能够很好地解释转录。

转录是基因表达的第一步。转录的产物包括 mRNA、rRNA 或 tRNA。

Transcription（转录）is the first step leading to gene expression. The stretch（伸展）of DNA transcribed into an RNA molecule is called a *transcription unit* and encodes at least one gene. If the gene transcribed encodes a protein, the result of transcription is messenger RNA（mRNA）, which will then be used to create that protein via the process of translation（翻译）. Alternatively, the transcribed gene may encode for either ribosomal RNA（rRNA）or transfer RNA（tRNA）, other components of the protein-assembly process（蛋白质组装过程）, or other ribozymes（核酶）.

5′非翻译区。
3′非翻译区。

A DNA transcription unit encoding for a protein contains not only the sequence that will eventually be directly translated into the protein（the coding sequence）but also regulatory sequences（调控序列）that direct and regulate the synthesis of that protein. The regulatory sequence before（upstream from）the coding sequence is called the five prime untranslated region（5′UTR）, and the sequence following（downstream from）the coding sequence is called the three prime untranslated region（3′UTR）.

转录具有一定的校对机制，但是转录比复制保真度低。

Transcription has some proofreading mechanisms, but they are fewer and less effective than the controls for copying DNA; therefore, transcription has a lower copying fidelity（保真度）than DNA replication.

DNA 按 3′→5′方向被转录。

As in DNA replication（复制）, DNA is read from $3' \to 5'$ during transcription（转录）. Meanwhile, the complementary（互补的）RNA is created from the $5' \to 3'$ direction. This means its 5′ end is created first in base pairing. Although DNA is arranged as two antiparallel（反向平行）strands in a double helix, only one of the two DNA strands, called the template strand（模板链）, is used for transcription. This is because RNA is only single-stranded, as opposed to double-stranded DNA.

DNA 双链中只有一条链用于转录称为模板链，另一条链则称为编码链，后者的序列与新生成的 RNA 转录本相同。

The other DNA strand is called the coding strand（编码链）, because its sequence is the same as the newly created RNA transcript（except for the substitution of uracil for thymine）. The use of only the $3' \to 5'$ strand eliminates the need for the

Okazaki fragments(冈崎片段) seen in DNA replication.

Transcription is divided into 5 stages: pre-initiation, initiation, promoter clearance, elongation and termination.

2.1.1 Pre-initiation 预起始

In eukaryotes(真核生物), RNA polymerase(RNA 聚合酶), and therefore the initiation of transcription, requires the presence of a core promoter sequence(核心启动子序列) in the DNA. Promoters are regions of DNA that promote transcription and, in eukaryotes, are found at -30, -75, and -90 base pairs upstream from the start site of transcription. Core promoters are sequences within the promoter that are essential for transcription initiation. RNA polymerase is able to bind to core promoters in the presence of various specific transcription factors(转录因子).

The most common type of core promoter in eukaryotes is a short DNA sequence known as a TATA box, found $25 \sim 30$ base pairs upstream from the start site of transcription. The TATA box, as a core promoter, is the binding site for a transcription factor known as TATA-binding protein (TBP), which is itself a subunit of another transcription factor, called Transcription Factor II D (TFIID). After TFIID binds to the TATA box via the TBP, five more transcription factors and RNA polymerase combine around the TATA box in a series of stages to form a preinitiation complex(预起始复合物). One transcription factor, DNA helicase(解螺旋酶), has helicase activity and so is involved in the separating of opposing strands of double-stranded DNA to provide access to a single-stranded DNA template(模板). However, only a low, or basal, rate of transcription is driven by the preinitiation complex alone. Other proteins known as activators and repressors, along with any associated coactivators(共激活因子) or corepressors(辅阻遏物), are responsible for modulating transcription rate.

Thus, preinitiation complex contains: Core Promoter

预起始复合物包含五种物质,即核心启动子序列、转录因子、DNA 解旋酶、RNA 聚合酶、激活因子与阻遏物。

Sequence、Transcription Factors、DNA Helicase、RNA Polymerase、Activators and Repressors. The transcription preinitiation in archaea(古生菌) is, in essence, homologous(同源的) to that of eukaryotes(真核生物), but is much less complex. The archaeal preinitiation complex assembles at a TATA-box binding site; however, in archaea(古生菌), this complex is composed of only RNA polymerase II, TBP(TATA 盒结合蛋白), and TFB (the archaeal homologue of eukaryotic transcription factor II B (TFIIB)).

2.1.2 Initiation 起始

细菌 RNA 聚合酶是由五种亚基组成的核心酶:2 个 α 亚基、1 个 β 亚基、1 个 β' 亚基和 1 个 ω 亚基。

In bacteria(细菌), transcription begins with the binding of RNA polymerase to the promoter in DNA. RNA polymerase is a core enzyme consisting of five subunits: 2 α subunits, 1 β subunit, 1 β' subunit, and 1 ω subunit. At the start of initiation, the core enzyme is associated with a sigma factor(σ 因子) that aids in finding the appropriate -35 and -10 base pairs downstream of promoter sequences.

真核生物转录起始更为复杂,真核生物的 RNA 聚合酶不直接识别核心启动子序列。

Transcription initiation is more complex in eukaryotes(真核生物). Eukaryotic RNA polymerase does not directly recognize the core promoter sequences. Instead, a collection of proteins called transcription factors(转录因子) mediate the binding of RNA polymerase and the initiation of transcription. Only after certain transcription factors are attached to the promoter does the RNA polymerase(RNA 聚合酶) bind to it.

在转录因子的介导下,RNA 聚合酶与启动子结合,三者共同形成转录起始复合物。

The completed assembly(组装) of transcription factors and RNA polymerase bind to the promoter, forming a transcription initiation complex. Transcription in the archaea(古生菌) domain is similar to transcription in eukaryotes.

2.1.3 Promoter clearance 启动子清除

After the first bond is synthesized, the RNA polymerase (RNA 聚合酶) must clear the promoter. During this time there is a tendency(倾向) to release the RNA transcript and produce truncated(缩短了的) transcripts. This is called abortive initiation(无效起始) and is common for both eukaryotes(真核生物) and prokaryotes(原核生物). Abortive initiation continues to occur until the σ factor rearranges(重新排列) resulting in the transcription elongation complex(转录延伸复合物) (which gives a 35 bp moving footprint). The σ factor(σ因子) is released before 80 nucleotides(核苷酸) of mRNA are synthesized. Once the transcript reaches approximately 23 nucleotides, it no longer slips and elongation can occur. This, like most of the remainder(剩余部分) of transcription, is an energy-dependent process, consuming adenosine triphosphate (ATP).

> σ因子的作用。
>
> 当转录本达到约23个核苷酸长度时,转录将进入延伸阶段。

Promoter clearance coincides with phosphorylation(磷酸化作用) of serine(丝氨酸) 5 on the carboxyl terminal domain(羧基端结构域) of RNA Pol(聚合酶) in eukaryotes(真核生物), which is phosphorylated by transcription factor II human (TFIIH).

> 转录因子 IIH(TFIIH)通过磷酸化作用,将真核生物 RNA 聚合酶羧基端结构域的丝氨酸5磷酸化,清除启动子。

2.1.4 Elongation 延伸

One strand of the DNA, the template(模板) strand (or noncoding strand), is used as a template for RNA synthesis. As transcription proceeds, RNA polymerase(RNA 聚合酶) traverses(通过) the template strand and uses base pairing complementarity with the DNA template to create an RNA copy. Although RNA polymerase traverses the template strand from $3' \to 5'$, the coding (non-template) strand and newly-formed RNA can also be used as reference points(参考点), so transcription can be described as occurring $5' \to 3'$. This produces an RNA molecule from $5' \to 3'$, an exact copy of the coding strand (except that thymines(胸腺嘧啶) are replaced with uracils(尿嘧啶), and the nucleotides(核苷酸) are composed of a ribose (5-carbon) sugar where DNA has deoxyribose(脱氧核糖) (one fewer oxygen atom) in its sugar-phosphate backbone).

> 转录生成的 RNA 延伸方向是 $5' \to 3'$。
>
> 转录本 RNA 与编码链的区别。

Simple diagram of transcription initiation.
RNAP＝RNA polymerase.

RNAP：RNA 聚合酶。

mRNA 转录与 DNA 复制不同。

Unlike DNA replication(复制), mRNA transcription can involve multiple RNA polymerases(RNA 聚合酶) on a single DNA template and multiple rounds of transcription (amplification of particular mRNA), so many mRNA molecules can be rapidly produced from a single copy of a gene.

延伸过程也包含校对机制。

Elongation also involves a proofreading mechanism that can replace incorrectly incorporated bases. In eukaryotes(真核生物), this may correspond with short pauses during transcription that allow appropriate RNA editing factors to bind. These pauses may be intrinsic(内在的) to the RNA polymerase or due to chromatin structure(染色质结构).

2.1.5 Termination 终止

细菌有两种不同的方式终止转录：不依赖 Rho 因子和依赖 Rho 因子方式。

Bacteria（细菌）use two different strategies for transcription termination. In Rho-independent transcription termination, RNA transcription stops when the newly synthesized RNA molecule forms a G-C-rich hairpin loop(发卡环) followed by a run of Us. When the hairpin forms, the mechanical stress breaks the weak rU-dA bonds, now filling the DNA-RNA hybrid(杂交体). This pulls the poly-U transcript out of the active site(活性中心) of the RNA polymerase(RNA 聚合酶), in effect, terminating transcription. In the "Rho-dependent" type of termination, a protein factor called "Rho" destabilizes the interaction between the template(模板) and the mRNA, thus releasing the newly synthesized mRNA from the elongation complex(延伸复合物).

微课：原核生物的蛋白质生物合成

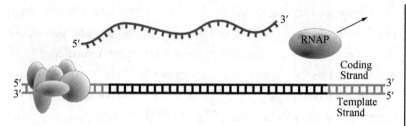

Simple diagram of transcription termination.

RNAP：RNA 聚合酶。

Transcription termination in eukaryotes（真核生物）is less understood but involves cleavage of the new transcript followed by template-independent（不依赖模板的）addition of as at its new 3′ end, in a process called polyadenylation（多聚腺苷酸化）.

真核生物的转录终止比较复杂，还没有完全搞清楚。

2.2 Reverse transcription 逆转录

Some viruses（病毒）(such as HIV, the cause of AIDS), have the ability to transcribe（转录）RNA into DNA. HIV has an RNA genome（基因组）that is duplicated into DNA. The resulting DNA can be merged with the DNA genome of the host cell（宿主细胞）. The main enzyme responsible for synthesis of DNA from an RNA template is called reverse transcriptase（逆转录酶）. In the case of HIV, reverse transcriptase is responsible for synthesizing a complementary DNA strand（互补的 DNA 链）(cDNA) to the viral RNA genome（病毒的 RNA 基因组）. An associated enzyme, ribonuclease H, digests the RNA strand, and reverse transcriptase synthesizes a complementary strand of DNA to form a double helix DNA structure（DNA 双螺旋结构）. This cDNA is integrated（整合）into the host cell's genome via another enzyme (integrase) causing the host cell to generate viral proteins that reassemble into new viral particles. In HIV, subsequent to this, the host cell（宿主细胞）undergoes programmed cell death, apoptosis（细胞凋亡）of T cells. However, in other retroviruses（逆转录病毒）, the host cell remains intact as the virus buds out of the cell.

逆转录的含义即有些病毒可以将 RNA 转录为 DNA。

逆转录酶的作用。

举例说明 HIV 逆转录的过程。

HIV 与其他逆转录病毒的不同之处，在于其宿主 T 细胞发生程序性死亡，而其他逆转录病毒宿主细胞保持完整的结构。

端粒酶的含义。	Some eukaryotic cells(真核细胞) contain an enzyme with reverse transcription(逆转录) activity called telomerase(端粒酶). Telomerase is a reverse transcriptase that lengthens(使延长) the ends of linear chromosomes(染色体). Telomerase carries an RNA template(模板) from which it synthesizes DNA repeating sequence, or "junk" DNA(垃圾DNA). This repeated sequence of DNA is important because, every time a linear chromosome(线性染色体) is duplicated, it is shortened in length. With "junk" DNA at the ends of chromosomes(染色体), the shortening eliminates some of the non-essential, repeated sequence rather than the protein-encoding DNA sequence farther away from the chromosome end-caution, not entirely true, indicates distortion of facts and bias(扭曲事实和偏见). Telomerase(端粒酶) is often activated in cancer cells to enable cancer cells to duplicate their genomes(基因组) indefinitely(无限制地) without losing important protein-coding DNA sequence. Activation of telomerase could be part of the process that allows cancer cells(癌细胞) to become immortal(永生). However, the true in vivo significance of telomerase has still not been empirically proven.
端粒酶的作用机制。	
端粒酶的真正意义还没有被证明。	

2.3 mRNA processing 信使RNA加工

真核生物初级RNA转录本需经过加工才能成为成熟的RNA。	Post-transcriptional modification(转录后修饰) or co-transcriptional modification(共转录修饰) is a process in cell biology by which, in eukaryotic cells(真核细胞), primary transcript RNA(初级RNA转录本) is converted into mature RNA. A notable example is the conversion of precursor(前体) messenger RNA into mature(成熟) messenger RNA (mRNA), which includes splicing(剪接) and occurs prior to protein synthesis. This process is vital for the correct translation of the genomes of eukaryotes, including humans, because the primary RNA transcript that is produced, as a result of transcription, contains both exons(外显子), which are either coding sections of the transcript or are important sequences involved in translation, and introns(内含子), which are the non-coding

sections of the primary RNA transcript.

The pre-mRNA molecule (mRNA 前体) undergoes three main modifications. These modifications are 5′ capping (加帽), 3′ polyadenylation (多聚腺苷酸化), and RNA splicing (剪接), which occur in the cell nucleus (细胞核) before the RNA is translated (翻译).

mRNA 前体在翻译前需要在细胞核中经过三个主要的修饰，包括 5′ 端加帽，3′ 端多聚腺苷酸化，以及 RNA 剪接过程。

2.3.1 5′ Processing 5′端加工

Capping of the pre-mRNA involves the addition of 7-methylguanosine (m7G, 7-甲基鸟苷) to the 5′ end. To achieve this, the terminal 5′ phosphate (磷酸) requires removal, which is done with the aid of a phosphatase enzyme. The enzyme guanosyl transferase (鸟嘌呤转移酶) then catalyzes the reaction, which produces the diphosphate (二磷酸) 5′ end. The diphosphate 5′ end then attacks the alpha phosphorus atom (磷原子) of a GTP molecule in order to add the guanine residue (鸟嘌呤残基) in a 5′5′ triphosphate (三磷酸) link. The enzyme (guanine-N7-)-methyltransferase (甲基转移酶) ("cap MTase") transfers a methyl group (甲基) from S-adenosyl methionine (S-腺蛋氨酸) to the guanine ring (鸟嘌呤环). This type of cap, with just the (m7G) in position is called a cap 0 structure. The ribose (核糖) of the adjacent nucleotide may also be methylated (甲基化) to give a cap 1. Methylation of nucleotides downstream of the RNA molecule produce cap 2, cap 3 structures and so on. In these cases the methyl groups are added to the 2′ OH groups of the ribose sugar. The cap protects the 5′ end of the primary RNA transcript from attack by ribonucleases (核糖核酸酶) that have specificity to the 3′5′ phosphodiester bonds (磷酸二酯键).

mRNA 前体加帽包含在 5′ 端添加 7-甲基鸟苷的过程。

解释如何在 5′ 端添加 7-甲基鸟苷的过程。

加帽的意义在于防止初级 RNA 转录本 5′ 端的 3′5′ 磷酸二酯键被特异性的核糖核酸酶攻击破坏。

2.3.2 3′ Processing 3′端加工

The pre-mRNA processing at the 3′ end of the RNA molecule involves cleavage (裂解) of its 3′ end and then the addition of about 250 adenine residues (腺苷酸残基) to form a poly(A) tail. The cleavage and adenylation (腺苷酸化)

mRNA 前体 3′ 端加工是指 3′ 端裂解，加入约 250 个腺苷酸残基，形成一个多聚腺苷酸尾巴。

reactions occur if a polyadenylation(多聚腺苷酸化) signal sequence (5′- AAUAAA - 3′) is located near the 3′ end of the pre-mRNA molecule, which is followed by another sequence, which is usually (5′-CA-3′) and is the site of cleavage. A GU-rich sequence is also usually present further downstream on the pre-mRNA molecule. After the synthesis of the sequence elements, two multisubunit(多亚基) proteins called cleavage and polyadenylation specificity factor (CPSF) and cleavage stimulation factor (CSF) are transferred from RNA polymerase II to the RNA molecule. The two factors bind to the sequence elements. A protein complex forms that contains additional cleavage factors and the enzyme polyadenylate polymerase (PAP,聚腺苷酸聚合酶). This complex cleaves the RNA between the polyadenylation sequence and the GU-rich sequence at the cleavage site marked by the (5′- CA - 3′) sequences. Poly(A) polymerase then adds about 200 adenine units to the new 3′ end of the RNA molecule using ATP as a precursor. As the poly(A) tail is synthesized, it binds multiple copies of poly (A) binding protein, which protects the 3′ end from ribonuclease(核糖核酸酶) digestion.

2.3.3 RNA splicing RNA 剪接

RNA splicing is the process by which introns(内含子), regions of RNA that do not code for protein, are removed from the pre-mRNA and the remaining exons(外显子) connected to re-form a single continuous molecule. Although most RNA splicing occurs after the complete synthesis and end-capping of the pre-mRNA, transcripts with many exons can be spliced co-transcriptionally(与转录同时地). The splicing reaction is catalyzed by a large protein complex called the spliceosome(剪接体) assembled from proteins and small nuclear RNA(小核RNA) molecules that recognize splice sites in the pre-mRNA sequence. Many pre-mRNAs, including those encoding antibodies, can be spliced in multiple ways to produce different mature mRNAs that encode different protein sequences. This process is known as alternative splicing(可变剪接), and allows production of a large variety of proteins from a limited amount of DNA.

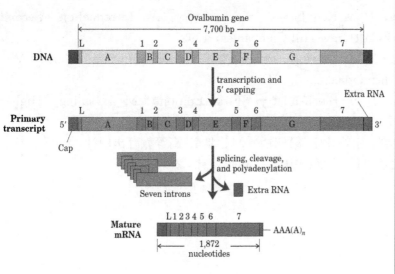

Overview of the processing of a eukaryotic mRNA.

【Key words】

transcription 转录	post-transcriptional modification 转录后修饰
reverse transcription 逆转录	preinitiation complex 预起始复合物
promoter 启动子	RNA polymerase RNA 聚合酶
exons 外显子	7-methylguanosine (m7G) 7-甲基鸟苷
introns 内含子	transcription factors 转录因子
eukaryote 真核生物	polyadenylation 多聚腺苷酸化
prokaryote 原核生物	telomerase 端粒酶
elongation 延伸	ribonuclease 核糖核酸酶
capping 加帽	alternative splicing 可变剪接

Questions

1. What is transcription? Please list the name of its 5 stages.

2. During transcription, which stand is the template strand and which strand is the coding strand?

3. Describe the RNA polymerase holoenzyme of *E. coli*.

4. Three main modifications occur in the pre-mRNA molecule before the RNA is translated, what are they?

References

[1] McKee Trudy, R McKee James. Biochemistry: An Introduction (Second Edition)[M]. New York: McGraw-Hill Companies, 1999.

[2] Nelson D L, Cox M M. Lehninger Principles of Biochemistry (Third Edition)[M]. Derbyshire: Worth Publishers, 2000.

[3] Ding W, Jia H T. Biochemistry (First Edition)[M]. Beijing: Higher Education Press, 2012.

[4] 郑集, 陈钧辉. 普通生物化学(第四版)[M]. 北京: 高等教育出版社, 2007.

[5] 王艳萍. 生物化学[M]. 北京: 中国轻工业出版社, 2013.

Chapter 3 Protein biosynthesis, modifications and degradation
蛋白质生物合成、修饰与降解

3.1 Translation 翻译

In molecular biology and genetics(遗传学), translation is the third stage of protein biosynthesis(生物合成)(part of the overall process of gene expression). In translation, messenger RNA (mRNA) produced by transcription(转录) is decoded by the ribosome(核糖体) to produce a specific amino acid chain, or polypeptide(多肽链), that will later fold into(折叠成) an active protein. In bacteria(细菌), translation occurs in the cell's cytoplasm(细胞质), where the large and small subunits(亚基) of the ribosome are located, and bind to the mRNA. In Eukaryotes(真核生物), translation occurs across the membrane(膜) of the endoplasmic reticulum(内质网) in a process called vectorial synthesis(矢量合成). The ribosome facilitates(促进) decoding by inducing the binding of tRNAs with complementary(互补) anticodon(反密码子) sequences to that of the mRNA. The tRNAs carry specific amino acids that are chained together into a polypeptide as the mRNA passes through and is "read" by the ribosome.

In many instances, the entire ribosome/mRNA complex will bind to the outer membrane of the rough endoplasmic reticulum(粗面内质网) and release the nascent(新生的) protein polypeptide inside for later vesicle(囊泡) transport and secretion(分泌) outside of the cell. Many types of transcribed RNA, such as transfer RNA, ribosomal RNA, and small nuclear RNA, do not undergo translation into proteins.

翻译是蛋白质生物合成的第三步。
由转录得到的信使 RNA 被核糖体解码产生特定的氨基酸链或多肽链,进而折叠成活性蛋白质。原核生物与真核生物翻译的区别。

核糖体通过诱导 tRNA 的反密码子与 mRNA 上与之互补的密码子结合而促进解码。
当 mRNA 经过核糖体时被其阅读,tRNA 则携带特定的氨基酸结合到多肽链上。

多数情况下,新生的蛋白多肽链会进入粗面内质网加工,进行囊泡运输并分泌到细胞外。有许多被转录的 RNA,例如转运 RNA、核糖体 RNA 以及小核 RNA,并没有被翻译成蛋白质。

翻译过程示意图。

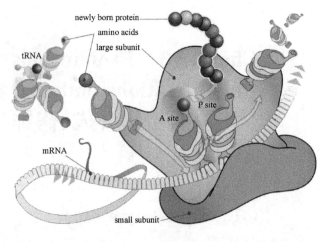

Diagram showing the translation of mRNA and the synthesis of proteins by a ribosome.

翻译包含四个阶段。

Translation proceeds in four phases: activation(活化), initiation(起始), elongation(延伸) and termination(终止)(all describing the growth of the amino acid chain, or polypeptide that is the product of translation). Amino acids are brought to ribosomes and assembled(组装) into proteins.

在活化阶段,氨基酸跟与之相应的转运RNA共价结合。氨基酸的羧基与转运RNA的3′端羟基通过酯键结合,结合有氨基酸的tRNA被称为"带电"。翻译起始时,核糖体的小亚基在起始因子的辅助下结合到mRNA的5′端。当核糖体A位遇到终止密码子(UAA, UAG 或UGA)时,翻译终止。

In activation, the correct amino acid is covalently(共价) bonded to the correct transfer RNA (tRNA). The amino acid is joined by its carboxyl group(羧基) to the 3′ OH of the tRNA by an ester bond (酯键). When the tRNA has an amino acid linked to it, it is termed "charged". Initiation involves the small subunit(亚基) of the ribosome binding to the 5′ end of mRNA with the help of initiation factors(起始因子, IF). Termination of the polypeptide happens when the A site of the ribosome faces a stop codon(终止密码子) (UAA, UAG, or UGA). No tRNA can recognize or bind to this codon. Instead, the stop codon induces the binding of a release factor (释放因子) protein that prompts the disassembly(拆卸) of the entire ribosome/mRNA complex.

许多抗生素通过抑制翻译发挥作用。真核生物与原核生物的核糖体结构不同,抗生素可以特异性地作用于细菌,而对宿主的真核细胞没有危害。

A number of antibiotics(抗生素) act by inhibiting translation; these include anisomycin(茴香霉素), cycloheximide(环己酰亚胺), chloramphenicol(氯霉素), tetracycline(四环素), streptomycin(链霉素), erythromycin(红霉素), and puromycin(嘌呤毒素), among others. Prokaryotic

(原核生物的) ribosomes have a different structure from that of eukaryotic(真核生物的) ribosomes, and thus antibiotics can specifically target bacterial infections(细菌感染) without any detriment(危害) to a eukaryotic host's cells.

3.1.1 Basic mechanisms 基本机制

The mRNA carries genetic information encoded as a ribonucleotide sequence(核苷酸序列) from the chromosomes (染色体) to the ribosomes(核糖体). The ribonucleotides are "read" by translational machinery in a sequence of nucleotide triplets(核苷酸三联体) called codons(密码子). Each of those triplets codes(三联密码子) for a specific amino acid.

密码子的概念。

The ribosome molecules translate this code to a specific sequence of amino acids. The ribosome is a multisubunit(多亚基) structure containing rRNA and proteins. It is the "factory" where amino acids are assembled(组装) into proteins. tRNAs are small noncoding(非编码) RNA chains (74～93 nucleotides) that transport amino acids to the ribosome. tRNAs have a site for amino acid attachment, and a site called an anticodon(反密码子). The anticodon is an RNA triplet complementary to the mRNA triplet that codes for their cargo(货物) amino acid.

核糖体是蛋白质合成场所。

tRNA 是氨基酸的载体，含有氨基酸臂和反密码子。

tRNA 结构示意图。

Tertiary structure of tRNA.

Aminoacyl tRNA synthetase(氨酰 tRNA 合成酶)(an enzyme) catalyzes the bonding between specific tRNAs and the amino acids that their anticodon sequences call for. The product of this reaction is an aminoacyl-tRNA(氨酰 tRNA) molecule. This aminoacyl-tRNA travels inside the ribosome, where mRNA codons are matched through complementary(互补的) base pairing to specific tRNA anticodons. The amino acids that the tRNAs carry are then used to assemble a protein. After the new amino acid is added to the chain, the energy provided by the hydrolysis(水解作用) of a GTP bound to the translocase(移位酶) EF-G (in prokaryotes) and eEF-2 (in eukaryotes) moves the ribosome down one codon towards the $3'$ end. The energy required for translation of proteins is significant. For a protein containing n amino acids, the number of high-energy phosphate bonds(高能磷酸键) required to translate it is $4n-1$. The rate of translation varies; it is significantly higher in prokaryotic cells (up to 17~21 amino acid residues per second) than in eukaryotic cells (up to 6~9 amino acid residues per second).

3.1.2 Structure prediction 结构推测

Whereas other aspects such as the 3D structure, called tertiary structure(三级结构), of protein can only be predicted using sophisticated(复杂的) algorithms(算法), the amino acid sequence, called primary structure, can be determined solely from the nucleic acid sequence with the aid of a translation table (转换表). This approach may not give the correct amino acid composition of the protein, in particular if unconventional(非传统的) amino acids such as selenocysteine(硒代半胱氨酸) are incorporated into the protein, which is coded for by a conventional stop codon in combination with a downstream hairpin (Selenocysteine Insertion Sequence, or SECIS)(硒代半胱氨酸插入序列).

There are many computer programs capable of translating a DNA/RNA sequence into a protein sequence. Normally this is performed using the Standard Genetic Code; many bioinformaticians(生物信息学家) have written at least one such program at some point in their education. However, few programs can handle all the "special" cases, such as the use of the alternative initiation codons. For instance, the rare alternative start codon CTG codes for methionine(蛋氨酸) when used as a start codon, and for leucine(亮氨酸) in all other positions.

	2nd base in codon				
1st base in codon	U	C	A	G	3rd base in codon
U	Phe Phe Leu Leu	Ser Ser Ser Ser	Tyr Tyr STOP STOP	Cys Cys STOP Trp	U C A G
C	Leu Leu Leu Leu	Pro Pro Pro Pro	His His Gln Gln	Arg Arg Arg Arg	U C A G
A	Ile Ile Ile Met	Thr Thr Thr Thr	Asn Asn Lys Lys	Ser Ser Arg Arg	U C A G
G	Val Val Val Val	Ala Ala Ala Ala	Asp Asp Glu Glu	Gly Gly Gly Gly	U C A G

Genetic code. The codon AUG both codes for methionine and serves as an initiation site: the first AUG in an mRNA's coding region is where translation into protein begins.

3.2 Posttranslational modification
翻译后修饰

Post-translational modification (PTM) refers to the covalent and generally enzymatic(酶的) modification of proteins during or after protein biosynthesis. Proteins are synthesized by ribosomes(核糖体) translating mRNA into polypeptide chains (多肽链), which may then undergo PTM to form the mature protein product. PTMs are important components in cell signaling(细胞信号).

Post-translational modifications can occur on the amino acid

翻译后修饰的各种方式及作用。	side chains(侧链) or at the protein's C- or N- termini(羧基或氨基端). They can extend the chemical repertoire of the 20 standard amino acids by introducing new functional groups such as phosphate(磷酸), acetate(醋酸), amide groups(酰胺基), or methyl groups(甲基). Phosphorylation(磷酸化) is a very common mechanism for regulating the activity of enzymes and is the most common post-translational modification. Many eukaryotic(真核的) proteins also have carbohydrate(糖) molecules attached to them in a process called glycosylation(糖基化), which can promote protein folding(折叠) and improve stability(稳定性) as well as serving regulatory functions. Attachment of lipid molecules, known as lipidation(脂化), often targets a protein or part of a protein to the cell membrane.
其他翻译后修饰的形式：肽段切除、二硫键形成。 举例：胰岛素。	Other forms of post-translational modification consist of cleaving(切开) peptide bonds(肽键), as in processing a propeptide(前肽) to a mature form or removing the initiator methionine residue(蛋氨酸残基). The formation of disulfide bonds(二硫键) from cysteine residues(半胱氨酸残基) may also be referred to as a post-translational modification. For instance, the peptide hormone insulin(胰岛素) is cut twice after disulfide bonds are formed, and a propeptide is removed from the middle of the chain; the resulting protein consists of two polypeptide chains connected by disulfide bonds.
有些翻译后修饰是氧化应激的结果，例如羰基化反应。	Some types of post-translational modification are consequences(结果) of oxidative stress(氧化应激). Carbonylation(羰基化反应) is one example that targets the modified protein for degradation(降解) and can result in the formation of protein aggregates(聚集). Specific amino acid modifications can be used as biomarkers indicating oxidative damage.
列举了常见的可以作为翻译后修饰位点的功能基团，这些基团在反应中作为亲核体。	Sites that often undergo post-translational modification are those that have a functional group(功能基团) that can serve as a nucleophile(亲核体) in the reaction: the hydroxyl groups(羟基) of serine(丝氨酸), threonine(苏氨酸), and tyrosine(酪氨酸); the amine(胺) forms of lysine(赖氨酸), arginine(精氨酸), and histidine(组氨酸); the thiolate anion(硫醇阴离子) of

cysteine(半胱氨酸); the carboxylates(羧基) of aspartate(天冬氨酸) and glutamate(谷氨酸); and the N- and C- termini. In addition, although the amides(酰胺) of asparagine(天冬酰胺) and glutamine(谷氨酰胺) are weak nucleophiles, both can serve as attachment points for glycans(多糖). Rarer modifications can occur at oxidized methionines(蛋氨酸) and at some methylenes(亚甲基) in side chains.

Post-translational modification of proteins can be experimentally detected by a variety of techniques, including mass spectrometry(质谱), Eastern blotting(印迹法), and Western blotting.

翻译后修饰可以通过一些实验技术来检测。

3.3 Protein degradation 蛋白质降解

Protein degradation may take place intracellularly(细胞内) or extracellularly(细胞外). In digestion of food, digestive enzymes(消化酶) may be released into the environment for extracellular digestion whereby proteolytic cleavage(蛋白酶切) breaks down proteins into smaller peptides(肽) and amino acids so that they may be absorbed(吸收) and used by an organism (生物体). In animals the food may be processed extracellularly in specialized digestive organs(器官) or guts, but in many bacteria(细菌) the food may be internalized(内化) into the cell via phagocytosis(吞噬作用). Microbial degradation(微生物降解) of protein in the environment can be regulated by nutrient availability(养分可用性). For example, limitation for major elements in proteins (carbon, nitrogen, and sulfur 硫) has been shown to induce proteolytic activity(蛋白水解活性) in the fungus (真菌) *Neurospora crassa* as well as in whole communities of soil organisms(土壤生物社区).

蛋白质降解可以发生在细胞内或细胞外。

食物中的蛋白质如何被消化吸收。

微生物降解环境中的蛋白质受到养分可用性的调节。

Proteins in cells are also constantly(经常) being broken down into amino acids. This intracellular degradation of protein serves a number of functions: It removes damaged and abnormal (异常的) protein and prevent their accumulation(堆积), and it

细胞中的蛋白质不断地被分解成氨基酸。细胞中的蛋白质降解具有多种功能。

also serves to regulate cellular processes by removing enzymes（酶）and regulatory proteins that are no longer needed. The amino acids may then be reused(重新使用) for protein synthesis.

细胞中的蛋白质降解可以通过两种途径实现：自溶酶体途径和泛素依赖途径。其中，自溶酶体途径一般不具有选择性。

The intracellular degradation of protein may be achieved in two ways: proteolysis(蛋白水解) in lysosome(溶酶体), or a ubiquitin-dependent(泛素依赖) process that targets unwanted proteins to proteasome(蛋白酶体). The autophagy-lysosomal (自噬溶酶体) pathway is normally a non-selective(非选择性) process, but it may become selective upon starvation(饥饿) whereby proteins with peptide sequence KFERQ or similar are selectively broken down. The lysosome(溶酶体) contains a large number of proteases(蛋白酶) such as cathepsins(组织蛋白酶).

泛素依赖途径通常具有选择性。

The ubiquitin-mediated process is usually selective(选择性的). Proteins marked for degradation are covalently linked to ubiquitin. Many molecules of ubiquitin may be linked in tandem(串联) to a protein destined for degradation. The polyubiquinated(多聚泛素化) protein is targeted to an ATP-dependent protease complex, the proteasome(蛋白酶体). The ubiquitin is released and reused, while the targeted protein is degraded.

Ubiquitin-proteasome system 泛素-蛋白酶体系统

蛋白酶体是普遍存在于真核生物和古生菌里的大分子蛋白复合体。蛋白酶体的主要作用降解细胞内不需要的或受到损伤的蛋白质，它参与调控特定蛋白质的浓度以及错误折叠的蛋白质降解过程。

Proteasomes(蛋白酶体) are very large protein complexes inside all eukaryotes and archaea(古生菌), and in some bacteria. In eukaryotes, they are located in the nucleus(核仁) and the cytoplasm(细胞质). The main function of the proteasome(蛋白酶体) is to degrade unneeded or damaged proteins by proteolysis, a chemical reaction that breaks peptide bonds. Enzymes that carry out such reactions are called proteases(蛋白酶). Proteasomes are part of a major mechanism by which cells regulate the concentration(浓度) of particular proteins and degrade misfolded(错误折叠的) proteins. The degradation process(加工) yields peptides of about seven to eight amino acids long, which can then be further degraded into amino acids and used in synthesizing(合成) new proteins.

Proteins are tagged(标记的) for degradation with a small protein called ubiquitin(泛素). The tagging reaction is catalyzed by enzymes called ubiquitin ligases(泛素连接酶). Once a protein is tagged with a single ubiquitin molecule, this is a signal to other ligases to attach additional ubiquitin molecules. The result is a polyubiquitin(多聚泛素) chain that is bound by the proteasome, allowing it to degrade the tagged(标记) protein.

泛素依赖的蛋白质降解过程。

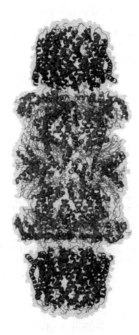

Cartoon representation of a proteasome. The caps on the ends regulate entry into the destruction chamber, where the protein is degraded.

In structure, the proteasome is a cylindrical(圆柱体) complex containing a "core"(核心) of four stacked(堆叠) rings around a central pore(气孔). Each ring is composed of seven individual proteins. The inner two rings are made of seven β subunits(亚基) that contain the six protease active sites(蛋白酶活性中心). These sites are located on the interior surface of the rings, so that the target protein must enter the central pore before it is degraded. The outer two rings each contain seven α subunits whose function is to maintain a "gate" through which proteins enter the barrel(圆筒). These α subunits are controlled

蛋白酶体的结构特点。

by binding to "cap" structures or regulatory particles that recognize polyubiquitin(多聚泛素) tags attached to protein substrates and initiate(启动) the degradation process. The overall system of ubiquitination and proteasomal degradation is known as the ubiquitin-proteasome system.

蛋白酶体降解途径的重要性。

The proteasomal degradation pathway is essential for many cellular processes, including the cell cycle(细胞周期), the regulation of gene expression, and responses to oxidative stress (氧化应激). The importance of proteolytic degradation inside cells and the role of ubiquitin in proteolytic pathways(蛋白水解途径) was acknowledged in the award of the 2004 Nobel Prize in Chemistry to Aaron Ciechanover, Avram Hershko and Irwin Rose.

【Key words】

ribosomes 核糖体	intracellularly 细胞内
codons 密码子	extracellularly 细胞外
anticodon 反密码子	phagocytosis 吞噬作用
aminoacyl tRNA 氨酰 tRNA	ubiquitin-dependent pathway 泛素依赖途径
multisubunit 多亚基	autophagy-lysosomal pathway 自噬溶酶体途径
posttranslational modification 翻译后修饰	proteasome 蛋白酶体
propeptide 前肽	polyubiquitin 多聚泛素
nucleophile 亲核体	misfolded proteins 错误折叠的蛋白质
degradation 降解	proteolytic 蛋白水解的

Questions

1. Describe briefly the process of translation in prokaryotes.
2. What is post-translational modification?
3. How about the intracellular degradation of protein?
4. Why antibiotics can kill bacteria by inhibiting translation, but does not harm the host's cells?

References

[1] McKee Trudy, R McKee James. Biochemistry: An Introduction (Second Edition)[M]. New York: McGraw-Hill Companies, 1999.

[2] Nelson D L, Cox M M. Lehninger Principles of Biochemistry (Third Edition)[M]. Derbyshire: Worth Publishers, 2000.

[3] Ding W, Jia H T. Biochemistry (First Edition)[M]. Beijing: Higher Education Press, 2012.

[4] 郑集,陈钧辉. 普通生物化学(第四版)[M]. 北京:高等教育出版社,2007.

[5] 王艳萍. 生物化学[M]. 北京:中国轻工业出版社,2013.